Energy Effectiveness

Sandra McCardell

Energy Effectiveness

Strategic Objectives, Energy and Water at the
Heart of Enterprise

 Springer

Sandra McCardell
Current-C Energy Systems, Inc.
Albuquerque, NM, USA

ISBN 978-3-030-07975-8   ISBN 978-3-319-90255-5   (eBook)
https://doi.org/10.1007/978-3-319-90255-5

Printed on acid-free paper

This Springer imprint is published by the registered company Springer International Publishing AG part of Springer Nature.
The registered company address is: Gewerbestrasse 11, 6330 Cham, Switzerland

Preface

Thomas Edison would be amazed at how much his so-called discovery of electricity has changed almost everything about the world and the way we operate in it; likewise, Rudolf Diesel would find it incredible that cars and trucks as well as ships, airplanes, and motorcycles run on what is called diesel fuel – and he would be astonished to know that diesel is made from petroleum, not the plant oil he developed.[1]

The world we know runs on energy. Those who live where there are frequent brownouts or blackouts when power is unavailable for a period of time know how difficult it is to operate a business in such circumstances; those who have lived through disasters with lengthy power outages know what it takes to do without the basic systems on which we have come to depend. Nevertheless, most of us take the energy upon which we rely for granted.

Energy is something we cannot see in use, yet it is generated with fuels we mine, refine, and send thousands of miles through natural gas pipelines, over electrical transmission lines, and on fuel tankers. It can be the most important "material" used in a manufacturing process or the main determinant of staff comfort in an office. Energy can be hazardous; it can be lifesaving. The long-term trend in energy costs (and, in some cases, the short-term trend as well) is upward, and the infrastructure which delivers that energy to us is variously aging or underdeveloped.

In the business sector, competition between firms is increasing, and yet one of the most important expenses is often unmanaged and uncontrolled. In the government and nonprofit sectors, energy poverty is being recognized as an immense challenge; when communities or individuals have insufficient access to energy, economic opportunities and development are inhibited. The energy production and generation industries are generally recognized as key economic drivers, yet energy resources which are created by savings are often unseen or ignored. Whether it is used or wasted, energy is invisible.

Energy, however, is *real*, and it should be visible. Energy can be locally generated, seen, measured, managed, controlled, saved, and reused. Following in the

---

[1] Pacific Biodiesel (2017) http://www.biodiesel.com/biodiesel/history/. Accessed 7/15/2017.

tradition of financial analysis and process improvement, it is possible to treat energy as one of the inputs to a process or a service which can be managed. Traditional business and social science techniques have a role to play in this technical arena where they can improve operations and cash flow. Using financial techniques to analyze energy projects, including breakthroughs and their use, combined with social science to affect change is relatively new and immensely powerful. It relies, to some extent, on new measurement technologies and the Internet of Things (IoT), where tiny circuits communicate wirelessly and data is managed in the cloud.

As James Harrington said in his process improvement book "Measurement is the first step that leads to control and eventually to improvement. If you can't measure something, you can't understand it. If you can't understand it, you can't control it. If you can't control it, you can't improve it." [2]

With proven techniques and new tools, energy can be measured, understood, controlled, and recycled.

Although the terms "energy efficiency," "energy conservation," "renewable energy," "alternative energy," and "energy effectiveness" are defined in more detail in the Introduction, this book uses the following convention: Energy conservation consists of turning equipment off; energy efficiency relies on installing more efficient equipment; renewable energy includes energy produced on-site, and energy effectiveness and alternative energy can include all of the above. The most common term used is "energy effectiveness" which includes the concept of energy being important because of what it can achieve. Water is also conceptually included as part of "energy effectiveness," for two reasons. One is that water and energy are intimately connected to one another, since 30% of all water use is in the production of energy and 30% of all energy is used to deliver water where it is needed. The other reason is that similar processes are used to evaluate and improve water resource and energy management.

This book is written for those whose main responsibilities are managing organizations or departments and for those willing to consider the potential benefits of alternative energy improvements for their organizations. Starting with general information and then more specifically, the book suggests ways in which energy improvements should be considered, evaluated, incentivized, implemented, and measured.

The approach described is an outgrowth of the work done by Current-C Energy Systems, Inc., the business I founded in 1996. Numerous books, white papers, webinars, policy initiatives, and articles have focused on the challenges to implementing energy efficiency and alternative energy projects, including examinations of the barriers to such investment and ways to overcome them. Those challenges persist, but the benefits and opportunities presented by energy projects are of greater interest, because there are many reasons to invest in alternative energy projects from an ongoing business perspective. Few clients are interested in conversations about new

---

[2] H. James Harrington (1991) Business Process Improvement: The Breakthrough Strategy for Total Quality, Productivity, and Competitiveness McGraw Hill Professional, cited at Business & Economics. http://www.goodreads.com/quotes/tag/measurement.

cooling technologies, utility rebate programs, solar panel efficiency, or water reuse, but many are concerned with the day-to-day issues involved in managing their organizations, their departments, or their projects. This book aims to connect energy effectiveness to those day-to-day issues. In fact, national and international studies have shown that well-managed companies must and do focus on energy effectiveness; investments in energy efficiency can drop utility costs from 10% to 50% below national averages for what are called "deep energy retrofits,"[3] and energy generation with renewable energy systems can reduce both risks and costs. For most organizations, this is an opportunity to be ignored only at one's competitive peril.

This is not an engineering book. There are hundreds of excellent engineering textbooks, white papers, and manuals on the market because most practitioners of the art and science of energy effectiveness are engineers. Others involved in energy issues are focused on policy at all levels of government, and although policies are extremely important, they are outside the control of most businesses. This book is written neither for engineers nor for policy makers. This practical manual uses standard business practices and techniques and discoveries from the social and behavior sciences that lead to consideration, assessment, and installation of energy effective projects.

New technologies play a role too, with dramatic and continuing innovations in the Internet of Things enabling data collection at the level of a fan on a plastic molding machine or the hot plate in a convenience store. Software disseminates that data and connects different divisions in an organization, so that actions taken in one place are recognized and adjusted for in another.

These changes are both exciting and potentially disruptive; they are likely to continue and accelerate, and therefore this book focuses less on any specific technologies than on the reasons to consider energy projects, ways to evaluate them, and other considerations and techniques for evaluating and generating enthusiasm for energy projects. First and foremost, the book emphasizes the importance of looking at energy as a strategically and operationally critical component for any organization.

For implementation to succeed in an environment with competing resource needs and infrequent management focus, techniques developed by behavior economics practitioners to frame analyses and choices in ways that resonate with real humans can be useful. For example, hyperbolic discounting describes the human tendency to discount the likelihood of a future result in comparison to a current situation; this bias makes it difficult for managers to purchase a new motor which will reduce costs 40% as long as the current motor is limping along.

Current-C Energy Systems, Inc., developed an informal process for evaluating and implementing potential energy projects. This book codifies those processes and provides several flexible and useful tools informed by techniques and approaches drawn from new research. A variety of adapted stories or case studies highlight important points. My hope is that this book helps readers understand that energy is

---

[3] New Buildings Institute (2017) http://newbuildings.org/hubs/deep-energy-retrofits/. Accessed 7/15/2017.

central to the success of organizations of all sizes, and that energy projects and programs are not mysterious but that practical approaches can be used to consider and implement them, even when technologies seem to change by the week. Using the framework and techniques presented, readers may be able to close the "energy efficiency gap" between projects which *should* be done because they are financially and technically viable and those that *are*, so that their organizations become more competitive and sustainable into the future.

Energy, for as long as it has been available, has been central to economic endeavors. This was no less true of the Phoenician traders whose ships used the wind to advantage, the sawmills in British Columbia which used rivers to transport logs to mills where they were turned into lumber, or the donkeys used to thresh grain in Portugal, than it is of the automobile manufacturers whose robots rivet metal pieces to make doors or the family run gas station with electrically operated fuel pumps. Energy has been central to these endeavors – but has not been central to management. This book is written to help change that relationship, the conversation, and the outcomes.

Albuquerque, New Mexico, USA   Sandra McCardell

References

Harrington, H. J. (2016, 11 14). *Business process improvement: The breakthrough strategy for total quality, productivity, and competitiveness (1991),* McGraw Hill Professional. Retrieved from Business & Economics/Goodreads (cited at). http://www.goodreads.com/quotes/tag/measurement.

New Buildings Institute. (2017, 7 15). *Deep energy retrofits.* Retrieved from New Buildings Institute. http://newbuildings.org/hubs/deep-energy-retrofits.

Pacific Biodiesel. (2017, July 15). *Pacific biodiesel.* Retrieved from Pacific Biodiesel. http://www.biodiesel.com/biodiesel/history/.

Contents

Part IV   Introducing and Using the Strategic Energy Effectiveness Framework (SEE): Analysis

Part V   Introducing and Using the Strategic Energy Effectiveness Framework (SEE): Planning, Implementation, Adaptation

Chapter 1
Introduction

1.1 Consulting, Advising, and Related Fields of Expertise

Advice is everywhere, and some of it is even good advice. There are countless books about how businesses, nonprofits, or institutions in general operate, written for managers and aspiring managers. A large proportion are about how to improve organizations and/or individuals: how to be a leader, remain competitive, improve manufacturing processes, measure key performance indicators, build a culture based on values, cope with economic downturns, and increase productivity. There are even more consultants available to assist managers with these efforts. This book is not intended to delve deeply into the varied business – or organization-related disciplines – although some basic methodologies and techniques drawn from them are recommended to understand energy use and improve performance.

In the field of energy use, there are also thousands of books and white papers or presentations about energy and water conservation, energy efficiency, renewable energy projects and technologies, and specific technologies or products such as solar thermal systems, variable speed drives, or small-scale wind turbines. The white papers and presentations are often narrowly focused on particular situations such as using solar thermal systems for space heating or heat recovery in hotel laundries. There are also volumes on green building, process improvement, landscaping, human behavior, organizational change, sustainability, waste management, water resources, and other related issues. They are rich and varied fields, and they all touch to some extent on energy and water issues because energy and water permeate all enterprises. This book, however, is more general, examining practical considerations and methodologies which can be applied consistently across different types of facilities.

With the focus of this book on business or organizational directors, managers, or owners, considerations of federal, state, or local policy are largely absent as well. They can and do impact energy reliability, pricing, risk, and availability, but they are not managed inside the organization. Similarly, the challenges of

© Springer International Publishing AG, part of Springer Nature 2018
S. McCardell, *Energy Effectiveness*, https://doi.org/10.1007/978-3-319-90255-5_1

utility-scale energy generation and transmission are absent, because they are outside the potential control of managers. In other words, the natural gas which is refined two states away and sent through a pipeline to headquarters is of general interest for understanding the industry, but only once it enters the building can it be controlled. Transportation is also outside the book's scope, although the same framework could be used to evaluate the effectiveness of a company-owned delivery truck which supplies raw materials or drops customer orders off. Those aspects of energy production and use are often included in sustainability accounting or certification programs such as the GRI initiative[1], the US Green Building Council (USGBC) but not here. This book looks only at what is in or around the building, whether it be a manufacturing facility, a hotel, or a small strip mall. The approach is holistic, addressing the building, the environment outside and inside that building, the occupants, the processes which take place in that building, and the procedures which map required actions. Those are the elements which management can most easily understand, affect, and control.

Naturally, this book contains many terms of art which are particular to the disciplines of either business or energy and engineering. Some terms are used throughout the book, and they are categorized and described in this section for ease of reference. Other terms are used less consistently, and where appropriate, I have defined most of those words in the chapters in which they are used.

1.2   Definition of Terms Used Generally

There are several core concepts or terms which are used frequently throughout the book. Because they come from both the business/finance and energy/water fields, they are categorized into those two groups for ease of reference.

1.2.1   Business/Finance

Behavior Economics   The developing discipline of behavior economics identifies the predictable ways in which individuals and groups operate in ways contrary to their economic interests and in contravention of traditional economic theory.

Business Purpose or Value   These terms are used as part of the Strategic Framework to express the reason the entity exists and what is intended to achieve. For nonprofits, another term might be mission.

Key Performance Indicator or KPI   This is a term used in traditional business analysis and management which transfers easily to the energy field. It is the outcome

---

[1] Global Reporting Initiative. (2017). https://www.globalreporting.org/. Accessed 7/15/2017.

of an organizational analysis conducted to determine which few indicators best capture ongoing business results and provide a quick indication of issues or achievements. In the energy field, obviously, the KPI should relate energy and water use in a strategic way to the value or purpose as well as other significant inputs and outputs of the business and be actionable.

Life Cycle Costing, LCC This traditional financial analysis tool attempts to take into consideration all the financial costs and benefits for a particular project through the projected end of life. It is most useful when used to compare proposed alternative projects or uses for fund and captures the time value of money and values for relevant variables such as maintenance costs, energy prices, disposal or sale, required supplies, and others.

Operations and Maintenance This discipline or division of every enterprise represents a category of activity that can be important in either wasting energy/water or reducing that waste; it should be considered in any energy projects or programs.

Paradigm Shift This term, borrowed from the social sciences, describes a powerful change in the way an individual or a group views the world.

Projects and Programs These two terms are used to refer to the planned activities recommended in this book. Although they can be interchangeable terms, the convention used here is to use projects when referring to one-time or short-term activities which often require an investment and programs to longer-term activities that require time and effort but often less investment.

Return on Investment (ROI) This is the preferred alternative to the payback period analysis and is drawn from standard financial analysis where it is the most common approach to analyzing multiple projects or investments. ROI captures as many relevant variables as possible for the current situation and the proposed alternatives, comparing them in financial terms. Some organizations may also set a *hurdle rate*, which defines the level of return above which projects should be implemented; others may compare the ROI for alternative projects and choose the one with the highest projected return.

Sustainability This term refers to the concept that current needs (of individuals, organizations, the environment) should be addressed without compromising the ability to address future needs. The term itself is rarely used in this book, but the concept is one that should resonate with all owners and managers intending that their organization should continue to thrive.

Variance Analysis Another traditional financial management tool, variance analysis quantifies the effect of several variables on the profitability of a particular product or service independently of one another. It can be a powerful tool, when the potential or actual effect of separate variables is important.

1.2.2   Energy/Water

Alternative Energy   For the purposes of this book, this term is taken to include all alternatives to the traditional (e.g., fossil fuel based and developed within the past 100 years or so) energies used to provide power inside an organization. Those include energy conservation, water use reduction, and energy efficiency as well as renewable energies and technologies such as heat recapture and reuse. This term is used as short-hand for "all of the above."

Building Envelope   This is the portion of the building which segregates the conditioned space inside the building from the unconditioned space outside.

Cost of Delay   A decision to invest in an energy saving projects is usually taken to reduce costs. For this reason, delaying the decision may be more costly than making the investment. The cost of delay captures this concept.

Electricity Terminology   Multiple terms in the book relate to electricity; the most common ones are defined below:

- Watts – Standard unit of power representing power in a circuit.
- Kilowatts – 1000 Watts (KW).
- Megawatts – 1,000,000 Watts (MW).
- kWh – Standard unit of electricity use over time; total energy in kWh is calculated by multiplying total Watts by time in hours.
- Demand charge – The additional cost for high instantaneous use of power.
- Peak/Off-Peak demand or load - The period when electrical power is required at higher than average supply levels for a period of time. Represented as kWD.
- Ratchet clause – A provision where the monthly demand charge is based on the highest measured demand in a previous period.

Energy Conservation   The term was coined in the United States during the oil shock of the 1970s and envisions turning energy-using equipment off or down, conserving energy by not demanding or using it. People conserve by taking action to do so. The effect of energy conservation is to reduce the time period during which the equipment is running.

In the 1970s, this was really the only way to reduce energy use – turn the thermostat down, turn the lights off, and drive fewer miles. People were urged to conserve energy even if the results were uncomfortable.

Energy Effectiveness   This term, coined by the author, is used in the book to include energy conservation, energy efficiency, water use reduction, and renewable energy systems and approaches in ways which allow the organization to better accomplish its strategic objectives.

Energy Efficiency This term is related to equipment improvements which accomplish more while using less power. Over the past 40 years, many building products, equipment, and control systems have become more energy efficient. Energy efficiency improvements do not require ongoing human interaction as they are generated by technical improvements.

Energy Conservation or Energy Efficiency Measure These terms are used by utilities, energy service companies, and others to describe the installations or actions recommended or taken to reduce energy use; they most commonly require an investment.

Energy Productivity This is a relatively new term, used in some publications to describe the increased productivity of an organization or an industrial process accomplished by employing both energy conservation and energy efficiency strategies. It measures the output and quality of goods and services generated with a given set of inputs and when used in a macrosense is the inverse of the energy intensity of GDP, measured as a ratio of energy inputs to GDP.

Energy Management Energy management is the practice of understanding and managing both energy use and energy generation (where applicable) so that an organization becomes energy effective. Energy management can be either automated or manual or combine both approaches. As used in this book, it is a recommended process which is distinct from an energy management system. For clarity, the term human energy management is also used.

Energy Management System An energy management system (EMS) is a system which is made up of controls for equipment and processes, focusing particularly on the heating/ventilation/air conditioning (or HVAC) system. An EMS is generally considered an energy efficiency system, and they can be very effective.

Energy Star The US Environmental Protection Agency's "Energy Star" program was one of the first to rate the energy use of appliances comparatively. Although "Energy Star" is not strictly speaking a term to be defined, it is frequently and is therefore included as if it were a term. The Agency has applied that same approach to evaluating buildings and has an extensive database of energy data organized geographically, by industry, etc. Many US-based studies rely on Energy Star information. Internationally, other governments or certifying agencies develop and use comparative data for different purposes; the Energy Star process is used as a proxy for these multiple and varied purposes.

Energy Use Intensity This calculation allows comparisons between different types of energy and different industries and building types. It is calculated by dividing the total energy (of all types) consumed by the sq.ft. of the building.

Energy-Water Nexus This term describes the deep connection between energy and water generation or distribution and use. Approximately 30% of energy use is dedicated to delivering potable water, and approximately 30% of water use is dedicated to generating energy.

Grid Interconnect System This term describes an alternative energy system (such as wind or solar PV) which has no battery backup system but is connected to the electric grid. These systems must have an automatic power disconnect in case of a power outage and can therefore not be used for backup power.

Integrative Planning and Design Once a system-based analysis or approach is completed, planning and design for new or retrofit designs can be completed in an integrative way. The effect of this approach is that installation costs for new systems and ongoing costs can both be minimized, because the planning and design process takes into account each system and its effect on the others.

Life Cycle Cost Analysis, LCCA This relatively new analytical approach is often confused with the LCC in the business/finance section above. The LCCA is intended to capture the costs and benefits of a product at each stage of its existence, from materials production through to disposal. The LCCA quantifies the costs and benefits of the *cradle-to-cradle* business process (where each product can be reused at the end of its useful life) as compared to a business as usual scenario. The LCCA can be a useful tool, but because it also includes the assignment of costs to *externalities* such as the environmental cost of mining ore or producing cotton or the environmental and social costs of reuse and recycling at the product's end of life, it is outside of the scope of this book.

LEED This acronym stands for Leadership in Energy and Environmental Design, a distinction originally developed by the US Green Building Council and now adopted globally. Other certification approaches tackle the problem in different ways, and LEED is used in this book as a proxy for the varied green building assessment and certification approaches.

Natural Gas Terminology Multiple terms in the book relate to natural gas; the most common ones are defined below:

- Btu – British thermal units. The heat required to raise the temperature of 1 pound of water by 1 degree Fahrenheit.
- Ccf – 100 cubic feet.
- Mcf – 1000 cubic feet.
- MMBtu – 1,000,000 British thermal units.
- Therm – 100,000 Btu or 0.10 MMBtu.
- $ per Ccf divided by 1.037 equals $ per therm.

Negawatts can be seen as an energy source – the ability to capture a resource that is otherwise wasted. The term was originally used by the Rocky Mountain Institute to mean a negative megawatt, a megawatt of power saved by increasing efficiency or reducing consumption. Physicist Amory Lovins coined the term and introduced it in a speech in 1989 although the term started life as a typo: Lovins saw *megawatt* spelled with an "n" in a document he was reading and was struck by the potential of that typo as a useful concept[2].

Payback Period, Simple Payback Period, PB This simple calculation for making energy project decisions has come into common use and is included with most proposals and specifications. The calculation consists of dividing the cost of a project – say, $10,000 – by the monthly utility savings projected, say, $115. In this case, the calculation would be $10,000/115 = 87 months or 7.25 years. In enterprises with less financial savvy, this calculation would likely kill the project, even though that may be the wrong financial decision. The variables upon which a standard financial analysis would rest (including price change expectations, maintenance cost adjustments, and other factors) are not captured in the simple payback period analysis.

Productive Use This term captures the work energy sources are purchased to accomplish. It represents the portion of the utility purchased and delivered that is productive, that is not wasted.

Rejected Energy This is the obverse of productive energy, that proportion of what was purchased that was rejected from the system, usually as heat. Rejected energy is a challenge at all stages of energy use including generation, conversion, and inefficient designs. An alternative term is *energy waste*.

Renewable Energy Renewable energy, in the context of this book, is that which is generated on-site or in a way that can be managed as if it were on-site. It is renewable, local, and controllable. Technologies included under this definition include solar photovoltaic for electricity generation, solar thermal, wind, geothermal in the form of a ground source heat pump, waste-to-energy systems, and passive solar.

Systems Analysis or Approach One of the challenges of designing and implementing an energy effective system is that the entire complex of building and the organization are interconnected. A lighting system can usefully be adjusted depending on the daylight available and the occupants in the building, as well as for the time of day. Water use may vary widely by production run, and the cooling load depends not only on the temperature outside and the number of people in the building but also on what type of activity they are involved in, how many computers are on, and what type of lights are being used. A systems analysis or approach looks at these systems both independently and together. Computerized building models make this task much less complicated than manual calculations.

---

[2]Lovins, A. (1989). http://www.ccnr.org/amory.html. Accessed 12/8/2016.

Water Terminology Multiple terms in the book relate to water; the most common ones are defined below:

- CCF – 100 cubic feet (1 cubic foot of water = 7.48 gallons).
- kGal – 1000 gallons.
- Sewer charges – Charges assessed for disposing of water, calculated in different ways by different utilities. Frequently higher than water charges.
- Stormwater charges – Charges assessed for stormwater which flows into drains on the property.
- Fire service charges are sometimes assessed on the water bill.

1.3   Goals of This Book

The book presents a framework that can be used to evaluate and manage energy use in organizations small and large, generally those without an energy manager or a sustainability department who would likely have most of the requisite knowledge in-house. A great deal of detail is provided to assist those who manage larger organizations, but the same approach is applicable to small entities; for these enterprises, much of the detail can be ignored, while the methodologies and framework are followed in a general way. Many chapters include examples designed to highlight the particular issues being addressed; the stories are combined or adapted from experience and case studies.

The goals of the book are to:

- Demonstrate the centrality of energy management for organizational success.
- Enable managers to understand energy and water inside their enterprise, and provide the tools to implement a process to control them.
- Highlight the power of using business tools to understand and control energy costs and processes.
- Demonstrate the importance of including people in the process.
- Assist managers by providing a way to control costs which have to date been considered uncontrollable.
- Describe ways to measure and evaluate energy programs, so that their effects can be seen and integrated into organizational culture.
- Introduce an integrated framework which can be used to understand, implement, and evaluate energy projects.
- Lead people and organizations to act.

1.4   Introduction of the Strategic Energy Effectiveness Framework (SEE)

The framework upon which this book is based is a cyclical one, shown in generalized form below and described in detail in Section III. It starts with the organization's value or purpose, the touchstone for everything else the enterprise does. Utilities, which are of course outside the control of the organization, provide benefits and have associated risks. The SEE process connects those external processes to the internal organization (Fig. 1.1).

Fig. 1.1 The Strategic Energy Effectiveness Framework (SEE). (Strategic Energy Effectiveness Framework, SEE, © 2016–2018 Current-C Energy Systems, Inc. Used with Permission)

1.5   Organization and Structure

The book is organized into sections and chapters which can be read separately but are intended to inform and build upon one another. Those sections are:

Section I The Context
This section consists of four chapters. Chapter 2 describes the importance of energy and water in the world and in organizations, including the ways in which we rely on energy and water and their economic benefits. Chapter 3 surveys relevant theories of management and organizations. Chapter 4 provides a history of energy use, including the development of programs to save energy. Chapter 5 focuses on current context, describing the ways energy is produced and consumed through energy flows, followed by descriptions of different types of energy.

Section II Organizations and Energy
This section consists of four chapters. Chapter 6 is about energy flows and management practices in organizations, and Chap. 7 introduces the "Four Fields" of analysis and focuses more specifically on where energy and water are used. Chapter 8 examines "drivers, barriers, and opportunities" or reasons that energy projects are often considered, reasons they may not be, and reasons that they should be. Chapter 9 highlights financial, technology, and people-oriented tools which support the Framework.

Section III The Strategic Energy Effectiveness Framework
Building upon the context provided in the first 2 Sections, the 11 chapters in Section III develop and demonstrate the SEE Framework:

Part A – Strategy begins with Chap. 10, which describes energy and organization as part of a system, including relevant leverage points. Chapter 11 examines the benefits and risks of utilities, and Chap. 12 introduces utility bills and their components.

Part B – Analysis consists of Chap. 13, External Assessment; Chapter 14, Setting Goals; and Chap. 15, Internal Assessment.

Part C – Planning, implementation, and adaptation consists of Chap. 16, Developing an Energy Business Plan; Chapter 17 Plan Implementation and Management; and Chap. 18, Evaluation, Recognition, and Adaptation.

Part D – Putting It All Together is composed of Chap. 19, a detailed examination of the Strategic Framework in practice, and Chap. 20, Checklists for the Framework.

A list of resources can be found in Chap. 21.

References

Global Reporting Initiative. (2017, July 15). *Global reporting initiative home*. Retrieved from Global Reporting Initiative: https://www.globalreporting.org/.

Lovins, A. (1989). *The Negawatt Revolution - solving the CO2 problem*. Green energy conference, Montreal 1989, Keynote Address (p. http://www.ccnr.org/amory.html). Montreal: Canadian Coalition for Nuclear Responsibility. Retrieved from Canadian Coalition for Nuclear Responsibility.

Part I
The Context

Chapter 2
Energy Around the Globe

2.1 Introductory Comment

This book is about practical approaches to using energy and water in organizations effectively, a subject which is critical to the future of those enterprises, to local communities, and to the world overall. To understand why it is so important to use these critical resources in productive ways, it is useful to consider what energy and water accomplish and how and where they are produced and employed. Modern civilization and culture depend on energy and water, and they are so ubiquitous that they are rarely examined.

2.2 The Importance of Energy and Water

The ability to heat and cool buildings using fossil fuels buried below land and sea has dramatically expanded the regions of the world in which people can live comfortably, freeing us from the constraints of heat and humidity as well as wind and bone-chilling cold. The ability to tap water deep underground or to make freshwater from the ocean has enabled people to live and grow food in deserts. Billions of people are living on resources such as oceans and water in streams, natural gas and coal, petroleum, and all its by-products that have developed over and been stored for eons.

Where energy is abundant, reasonably priced, and reliable, economic activity increases. That situation prevails in many parts of the world, but where there are few gas pipelines, water lines, or electric grids, economic activity is more difficult to generate. And almost universally, whether there is currently sufficient infrastructure or not, that infrastructure is aging and causing concerns about reliability in the future.

© Springer International Publishing AG, part of Springer Nature 2018   15
S. McCardell, *Energy Effectiveness*, https://doi.org/10.1007/978-3-319-90255-5_2

Maps provide a visual way to understand energy around the world. In collecting them, I have focused on one commodity, electricity, rather than natural gas, petroleum products (such as diesel), or water; for those interested in learning more, the websites referenced have a great deal of information on those commodities as well as more information on electricity.

2.3   Energy and Economics Around the World

Figure 2.1 highlights electricity production by country, with the darker colors representing higher production. Several points stand out:

- China, as a large producer, contributes significantly to greenhouse gas emissions and has easy access to energy for factories, a contributor to their comparatively low cost of production.
- Several African countries produce little to no electricity and must therefore import it, increasing costs through all economic sectors. The alternative is to do without, putting their countries in a situation of energy poverty, where neither organizations nor individuals have access to energy for lighting, factories, or other uses. Without local energy resources, social and economic costs are high.

Comparing Fig. 2.1 on energy production to Fig. 2.2 on energy consumption, it becomes evident that the highest consuming countries tend to be in in North America, the Soviet Union, Western Europe, and other developed countries.

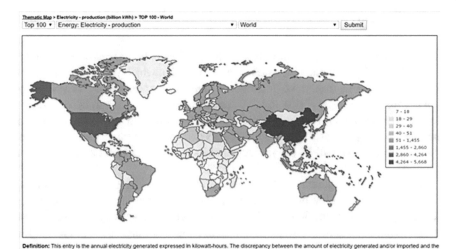

Fig. 2.1 Map of electricity production by country. (Mundi (2017). https://www.indexmundi.com/map/?t=100&v=79&r=xx&l=en. Accessed 8/3/2017)

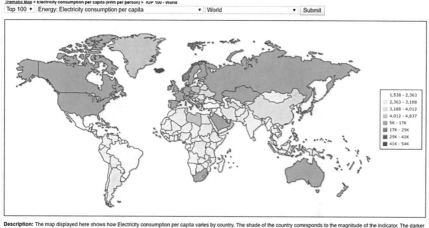

Fig. 2.2 Electricity consumption per capita. (Index Mundi (2017a). https://www.indexmundi. com/map/?t=100&v=81000&r=xx&l=en. Accessed 8/3/2017)

(A developed country is seen in this context as one with a developed energy infrastructure.) Some of these produce their own energy, and others import it from elsewhere – increasing both costs and security or reliability concerns.

- China is again an interesting case, with lower current consumption per capita than other industrialized countries. As the Chinese people and industries increase their energy requirements, total consumption is likely to increase dramatically.
- Mexico produces more electricity than it consumes, as do Brazil and several other countries in Latin America.
- There are many countries, especially island nations, which source most or all of their energy products from elsewhere.

The gross domestic product or GDP of a country, a standard measure of total economic activity, shows the influence of those energy production and use patterns. The information in Fig. 2.3, GDP per capita, can be combined with that in the two previous graphs to see the relationship between GDP and electricity.

- Australia, with relatively high GDP, also produces its own energy and has a high consumption level, some of which is related to manufacturing that increases GDP.
- There is a similar pattern in Western Europe, the United States, Saudi Arabia, and Canada.

Manufacturing might be considered a proxy for commercial and industrial facilities, a subset of the target readership for this book, so a more detailed focus on "manufacturing" is shown in Fig. 2.4.

- China and the United States stand out as high manufacturing output countries.
- Countries such as Afghanistan, Libya, and the Central African Republic have very little manufacturing output.

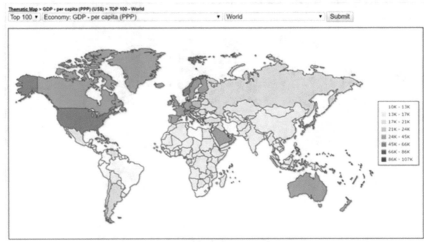

Fig. 2.3 GDP per capita. (Index Mundi (2017b). https://www.indexmundi.com/map/?t=100&v=67&r=xx&l=en. Accessed 8/2/2017)

Fig. 2.4 Map of manufacturing output. (NationMaster (2017a). NationMaster.com "Countries Compared by Industry > Manufacturing output. International Statistics at NationMaster.com," World Bank national accounts data, and OECD National Accounts data files. Aggregates compiled by NationMaster. Retrieved from http://www.nationmaster.com/country-info/stats/Industry/Manufacturing-output. Accessed 8/3/2017)

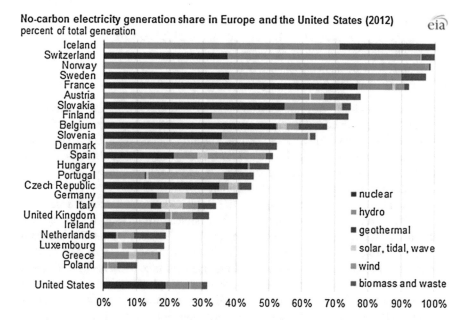

No-carbon electricity generation share in Europe and the United States (2012)
percent of total generation

Fig. 2.5 No-carbon electricity generation in Europe and the United States (2012). (Energy Information Administration (n.d.). https://www.eia.gov/todayinenergy/detail.php?id=18071. Accessed 8/7/2017)

Fossil fuels still predominate worldwide, but alternative energy generation is growing. Figure 2.5 shows the varied renewable energy generation patterns among countries in Europe as well as the United States. Iceland, for example, relies heavily on geothermal energy; Italy and Germany have the largest percentages of solar/tidal/wave generation, and France's nuclear component in 2012 was significant (Fig. 2.5).

Electricity is rarely generated where it is to be used, and the systems required to deliver electricity where it is to be used are inadequate in much of the world. Issues with rising costs to cover investments in this infrastructure are likely to cause increasing energy reliability challenges everywhere.

Figure 2.6 has two components: A map which visually shows a comparison by country of the number of days of electrical outages and selected datapoints upon which the map was based. In 2002, Bangladesh had 249 days of outages, and in 2005 Eastern European countries had 16. Operating an enterprise in either circumstance would be challenging.

Selected datapoints from the chart show that:

- Bangladesh had 249 days of outages in 2002.
- Albania had 194 days of outages in 2005.
- Kenya had 84 days of outages in 2003.
- India had 67 days of outages in 2006.

- Mexico had 24 days of outages in 2006.
- Average days of outages for Eastern Europe in 2005 were 16.
- The Philippines had 5 days of outages in 2003.
- Ireland had 1.66 days of outages in 2005.
- Germany had 0.23 days of outages in 2005.

In remote regions with few generation resources and few capital resources to invest in energy infrastructure, both energy access and energy security suffer; the economic and social consequences of this disparity can be significant. The World Economic Forum map in Fig. 2.7 highlights this disparity. Where there are low levels of energy access and security, businesses and individuals must rely on local power they themselves provide or do without.

One example of the complexity that develops with unequal energy access and security was on show during the 2015 discussions surrounding the importance, challenges, and agreements for the United Nations Climate Change Conference in Paris, also known as COP 21. Countries without sufficient energy to develop their economies felt that the richer countries were trying to hold them back, while those richer countries felt they were being unfairly penalized. Now that the Agreement is in force, the major differences between the industrialized countries where consistent, reliable energy supplies are a given and other countries with difficult energy access issues remain. That discussion is outside the scope of this book, but

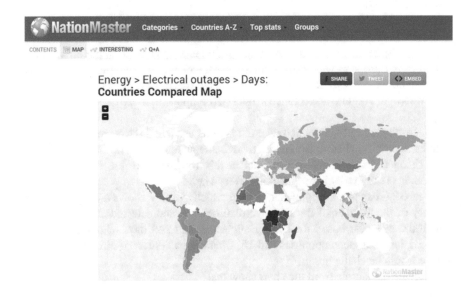

Fig. 2.6 Electric outages in days by country. (NationMaster (2017b). NationMaster.com http://www.nationmaster.com/country-info/stats/Energy/Electrical-outages/Days#. Accessed 8/3/2017)

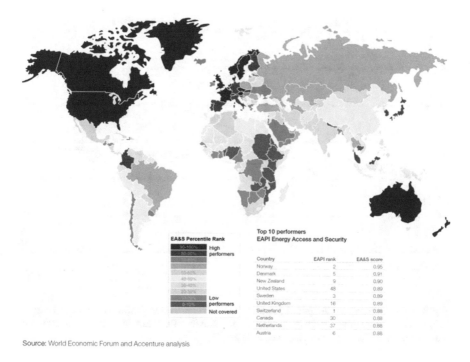

Fig. 2.7 Top 10 Performers / Energy Access and Security. (World Economic Forum (2017). http://reports.weforum.org/global-energy-architecture-performance-index-report-2016/energy-access-and-security/. Accessed 7/6/2017)

it is important to remember that reliable energy enables more broad-based economic development than was possible for most of human history.[1]

As nations grow, their demands for energy increase. The EIA, or US Energy Information Administration, states in its International Energy Outlook 2016[2] "Much of the world increase in energy demand occurs among the developing non-OECD nations (outside the Organization for Economic Cooperation and Development), where strong economic growth and expanding populations lead the increase in world energy use. Non-OECD demand for energy rises by 71% from 2012 to 2040. In contrast, in the more mature energy-consuming and slower-growing OECD economies, total energy use rises by only 18% from 2012 to 2040." This trend is shown in Fig. 2.8, also drawn from the report.

The more detailed projection by region in Fig. 2.9 shows significant growth – both actual and projected – particularly in Asia with its growing economies.

---

[1] In Brief, National Assembly for Wales Research Service. (2016). https://assemblyinbrief.wordpress.com/2016/04/04/the-paris-agreement-on-climate-change-a-summary/. Accessed 6/30/2017.
[2] US Energy Information Administration. (2016). International Energy Outlook 2016, Report Number DOE/EIA-0484(2016) https://www.eia.gov/outlooks/ieo/world.php. Accessed 5/20/2017.

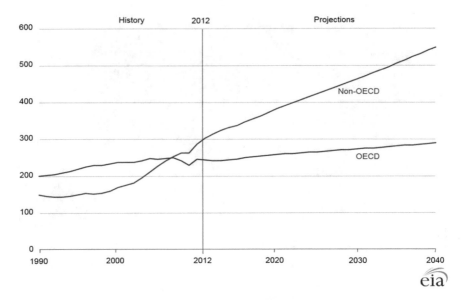

Fig. 2.8 World energy consumption by region 1990–2040. (US Energy Information Administration (2016), International Energy Outlook 2016, Report Number DOE/EIA-0484(2016) https://www. eia.gov/outlooks/ieo/world.php. Accessed 5/20/2017)

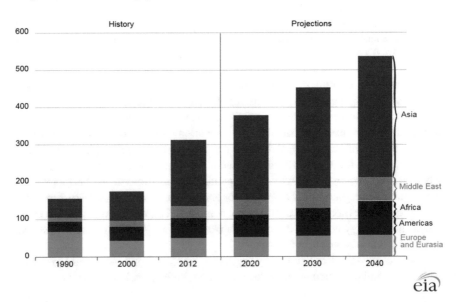

Fig. 2.9 Non-OECD energy consumption by region 1990–2040. (US Energy Information Administration (2016), International Energy Outlook 2016, Report Number DOE/EIA-0484(2016) https://www.eia.gov/outlooks/ieo/world.php. Accessed 5/20/2017)

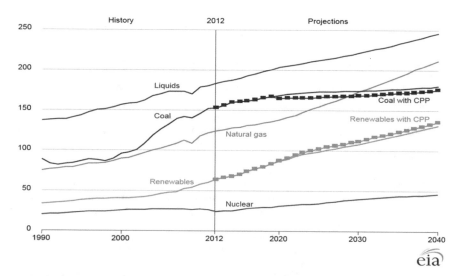

Fig. 2.10 Non-OECD energy consumption by energy source, 1990–2040. (US Energy Information Administration (2016), International Energy Outlook 2016, Report Number DOE/EIA-0484(2016) https://www.eia.gov/outlooks/ieo/world.php. Accessed 5/20/2017)

The mix of fuels is also projected to change, as shown in Fig. 2.10 (which does take into account country commitments under the COP and shows the projection under the US Clean Power Plan as originally proposed, as well as without it). Nuclear is shown as leveling off, along with coal; natural gas shows the steepest growth curve.

These diagrams highlight the importance of energy for economic activity, and project a future that continues to rely on energy, from multiple sources. For most organizations, the energy sources will remain the same, but they are likely to contend with increasing costs, price volatility, and reliability issues.

References

Alliance to Save Energy. (2013). *The history of energy productivity*. Washington DC: Alliance to Save Energy. Retrieved from http://www.ase.org/sites/ase.org/files/resources/Media%20 browser/ee_commission_history_report_2-1-13.pdf.

American Council for an Energy Efficient Economy. (2016). World energy efficiency scoreboard 2016. Retrieved from Aceee.org: http://aceee.org/sites/default/files/image/topics/2016-world-scores.png.

Cantore, N. (2014). Factors affecting the adoption of energy efficiency in the manufacturing sector of developing countries. *Energy Efficiency*. https://doi.org/10.1007/s12053-016-9474-3.

European Union. (n.d.). Googlegroups copied from EU. Retrieved from Energy Discussion: https://16625575041838918609.googlegroups.com/attach/f369274911 2f9638/EU27_06.png ?part=0.1&view=1&vt=ANaJVrFPHPNTdzkcLtEHOODI-aQfN0I9-tV0KlZI7GwmgTY-AAdxqZgdjhEbEkdxyCndXuF4RzKvR3NOxZ1eJunVuE0DCAtxY8Cvm3pyb31recP3Z-kY23zQ.

Fry, A., et al. (2005/Reprint 2006). Facts and Trends - Water. London UK: World Business Council for Sustainable Development, wbcsd.org.

Index Mundi. (2017a, August 3). Thematic Map - Electricity Consumpmtion per capita. Retrieved from Index Mundi: https://www.indexmundi.com/map/?t=100&v=81000&r=xx&l=en.

Index Mundi. (2017b, August 3). Thematic Map - GDP - Economy: GDP per capita. Retrieved from Index Mundi: https://www.indexmundi.com/map/?t=100&v=67&r=xx&l=en.

Laitner, J. A. (2012). *The long-term energy efficiency potential: What the evidence suggests, report number E121*. Washington, DC: American Council for an Energy-Efficient Economy.

National Assembly for Wales Research Service. (2016, April 4). The Paris agreement on climate change - a summary. Retrieved from Assembly in Brief: In Brief, National Assembly for Wales Research Service (2016) https://assemblyinbrief.wordpress.com/2016/04/04/the-paris-agreement-on-climate-change-a-summary/. Accessed 6/30/2017.

NationMaster. (2017a, August 3). Countries compared by industry - manufacturing output. International Statistics. World Bank national Accounts Data and OECD National Accounts Data Files. Retrieved from NationMaster.com: http://www.nationmaster.com/country-info/stats/Industry/Manufacturing-output.

NationMaster. (2017b, 8 3). country-info/stats/Energy/Electrical-outages/Days#. Retrieved from NationMaster.com: NationMaster.com http://www.nationmaster.com/country-info/stats/Energy/Electrical-outages/Days#.

Thematic Map - Population - World - Demographics: Population. (2017, August 3). Retrieved from Index Mundi: https://www.indexmundi.com/map/?t.

US Department of Energy, Halverson, et al. (2014). *ANSI/ASHRAE/IES standard 90.1–2013 determination of energy savings: Qualitative analysis PNNL-23481*. Richland: Pacific Northwest National Laboratory.

US Energy Information Administration. (2016). International Energy Outlook 2016, Report # DOE/EIA-0484(2016). Washington DC: US Energy Information Administration https://www.eia.gov/outlooks/ieo/world.php. Accessed 5/20/2017.

US Energy Information Agency. (n.d.). faqs. Retrieved from eia.gov: https://www.eia.gov/tools/faqs/faq.php?id=427&t=3.

Valentine, K. (2015, August 13). It's not a Pipe Dream: Clean energy from water pipes comes to Portland. Retrieved from Thinkprogress.org: https://thinkprogress.org/its-not-a-pipe-dream-clean-energy-from-water-pipes-comes-to-portland-16240cc33412/.

World Economic Forum. (2017). Reports. Retrieved from weforum.org: http://reports.weforum.org/global-energy-architecture-performance-index-report-2016/energy-access-and-security/. Accessed 7/6/2017.

Young, R., et al. (2014). *The 2014 international energy efficiency scorecard report E1402*. Washington DC: American Council for an Energy-Efficient Economy.

Chapter 3
Managing in Organizations

3.1 Technical Decisions, Management Decisions

Energy efficiency and renewable energy decisions are often presented as technical decisions, in technical terms. This approach is inappropriate in most business or organizational environments because strategic, financial, and personnel issues, along with technical concerns, should be part of the analysis. Decisions are better made in the context of management theory and decision-making processes, a brief history of is presented in this chapter.

Management theory is generally considered to have begun late in the nineteenth century, with rapid industrial growth that increased the size and complexity of manufacturing businesses beyond what could be easily managed by a small group. This growth brought the rise of professional managers who were not owners.[1]

3.2 Business Processes and Management Theory

The main approaches to business and process management theory have evolved over time, each one generally drawing upon several previous theories. In chronological order, some of the theories most relevant to energy use and management are summarized below. The theories have moved from linear approaches focusing on one type of influence or idea, to those which are based on a richer model of the organization as a system.

- "Scientific management" was promulgated by Frederick W. Taylor in the 1880s. He believed (as a mechanical engineer) that higher manufacturing productivity depended on designing jobs appropriately and providing production incentives, laying the foundation for later management systems.

---

[1] Bosman Readmore. (2009). http://faculty.wwu.edu/dunnc3/rprnts.historyofmanagementthought. pdf. Accessed 8/4/2017.

© Springer International Publishing AG, part of Springer Nature 2018
S. McCardell, *Energy Effectiveness*, https://doi.org/10.1007/978-3-319-90255-5_3

- About a decade later, Jules Henri Fayol, working for a French mining company, determined that management was both important and common to all human activities which require planning, organizing, commanding, coordinating, and controlling; he was convinced that management was a skill that could be taught.
- At around the same time in Germany, Max Weber focused on organizational structure, proposing clear lines of authority and control that helped large organizations to function in a more organized and stable manner as bureaucratic systems.
- At the beginning of the twentieth century, Mary Parker Follet added to the framework elements of human relations and structure; she was a forerunner of theories which put people and groups at the center of enterprises and also recognized the importance of factors external to the entity.
- Elton Mayo, at Harvard, also focused on the importance of individuals in the system as a result of the Hawthorne Studies in 1932. These appeared to demonstrate that participation in social groups and group pressure had the strongest impact on worker productivity.
- Kurt Lewin started the Research Center for Group Dynamics at MIT in 1946, becoming the "Father of Organizational Development."
- The ISO standard was organized in 1946 as the successor to a previous standards organization with the intention of providing clear references standards recognized internationally, based on quality and reliability, compatibility, and other standards. ISO has evolved in many directions; ISO 50001:2011 concerns energy management systems and is therefore the system most relevant to this book. The US Department of Energy notes on Energy.Gov that "Manufacturers, corporations, utilities, energy service companies, and other organizations are using ISO 50001 to reduce costs and carbon emissions. More than 7300 sites worldwide achieved ISO 50001 certification by May 2014—increasing 234% in just over a year."[2]
- In 1949, another mining operation study by London's Tavistock Institute of Human Relations determined that both social and technical attributes are important for business success.
- Quality assurance programs developed by Edward Deming in Japan after World War II to improve industrial equipment led to the concept of total quality management (TQM) in 1954 by Dr. Joseph Duran and Kaoru Ishikawa. The major elements of TQM include quality before short-term profits, customers ahead of the producers, customers as the next "process," fact-based decisions, participatory management, and cross-functional committees across all functions.
- Peter Drucker wrote in *The Practice of Management*, in 1954, that managers should focus on "What is our business and what should it be," designing the organizational structure to reach the business' goals over the short-term and

---

[2] 50001. (n.d.). https://energy.gov/ISO50001. Accessed 3/27/2017.

medium term. "Management by objectives" or MBO was one of Drucker's most creative concepts, modified by Schleh to "management by results."

- In 1978, Tom Gilbert wrote *Human Competence: Engineering Worthy Performance*. In this book, the focus was on accomplishments as the best targets for performance requirements as they are geared towards observable, specific measurements important to the organization.
- Also in 1978, Tom Peters and Robert H Waterman Jr. published *In Search of Excellence*, a book that continues to be widely read. The authors found eight themes they saw as responsible for the success of the world's most successful corporations. Peters noted in 2001 that the essential message of the book is that success comes from:

  - People
  - Customers
  - Action[3]

- Six Sigma focuses on process improvement and was developed by Bill Smith at Motorola before being adopted and adapted at General Electric under Jack Welch in 1995; hundreds of other companies and organizations have instituted Six Sigma processes since then. As a prescriptive method, Six Sigma specifies steps and targets for each project in a continuous process with the goal of achieving near zero defects.
- In 1990, Peter Senge wrote about the "learning organization", one which understands itself as a complex system with a vision and purpose, using feedback systems to achieve goals and relying on teams and leadership throughout the organization.
- LEAN management is credited to Toyota and described in *The Machine That Changed the World* by James Womack, Daniel Roos, and Daniel Jones in 1990. The principles of LEAN management include specifying what the customer wants, identifying value streams while challenging all the wasted steps and inputs, delivering continuous flow, and continuous improvement.

In the field of energy, there are fewer general theories and more protocols developed to solve a particular challenge. One of those with important impact on the energy field started in the 1990s and is called the IPMVP or International Performance Measurement and Verification Protocol, published by the Efficiency Valuation Organization or EVO[4]. That organization specifies methodologies for calculating savings, particularly important where contractors are being paid based on those savings or where funds are loaned to build projects based on savings projections. Because the IPMVP measures financial results from one or more simultaneous energy projects, it could be considered both a financial and an energy-related measure.

---

[3] Businessballs.com (2017). http://www.businessballs.com/tompetersinsearchofexcellence.htm.

[4] Efficiency Valuation Organization. (2009). evo-world.org/en/ or http://mnv.lbl.gov/keyMnV-Docs/ipmvp.

3.3   Contemporary Theories Related to Managing Energy

Contemporary management theories are broader than the classical norm and provide a more nuanced or complicated view of activities inside an enterprise. Utilities too are nuanced and complicated, sitting at the heart of enterprise as they do, and this similarity forms part of the base for the Strategic Framework.

Contemporary theories include the following:

- Contingency theory posits that situations and organizations are varied and fluid and that managers must consider all aspects of a situation so that their decisions account for those variables which are most relevant.
- Chaos theory disputes the previous deterministic theories that what happens in an organization is predictable and controllable, suggesting instead that systems move toward complexity and volatility.
- Systems theory (originally developed to explain complex living systems) has also been applied to organizations. All systems are composed of interrelated parts, and change in one area affects the others. In addition, influences from outside the system can have an effect. Systems thinking broadens management's perspective and helps to interpret patterns and initiate changes in the system.

The systems thinking approach is most closely aligned with the view of the SEE Framework which leans towards a systemic analysis and plan, a core focus on the importance of data, team leadership, and continuous improvement.

A detailed examination of these management theories is beyond the scope of this book but links to many of the theories and methodologies are at the end of this chapter.

3.4   The Five-Step Framework for Institutional Change

These different theories about leadership, organizational structure, and change are important because they provide ways to understand how and why energy-related decisions are made. Nicolas Baker at the US Department of Energy, Amy Wolfe at Oak Ridge National Laboratory, and Christopher Payne at Lawrence Berkeley National Laboratory have taken these theories further and developed a Five-Step Framework for Institutional Change as shown in Fig. 3.1. This Five-Step Framework was a precursor to the Strategic Energy Effectiveness (SEE) Framework, and the tools which form part of the Five-Step Framework are useful for human behavior programs; therefore a brief description is presented here and worksheets are included in Chap. 20.

Fig. 3.1 The Five-Step Framework. (Baker (2017). https://energy.gov/eere/femp/institutional-change-federal-energy-management)

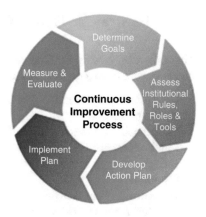

The Five-Step Framework for institutional change is a cycle, with the stages being:

1. Determine goals.
2. Assess institutional rules, roles, and tools.
3. Develop an action plan.
4. Implement the plan.
5. Measure and evaluate.

The power of the framework comes from combining organizational development and structure with sociology and anthropology and new theories about what motivates people to take action. The framework is a cycle which operates at different depths.

Once goals are determined, the next step is to assess the institutional context – how does this particular organization work, and how is it structured? Since institutional change must be system-wide to be effective, it is important to map the roles of key players and the impact they have and should have on the goal.

There are also rules to be defined, identified, and examined – some of which are formal and some of which are informal. Those too may need to change.

Tools are the next items to be considered – what are they and how do they impact the current situation and the desired future?

The action plan should provide opportunities to engage and educate the people in the organization, enable them to take action, and evaluate the results.

The eight principles to encourage change are social network, social empowerment, social commitment, information and feedback, multiple motivations, leadership, infrastructure, and continuous change and innovation.[5]

---

[5] Baker. (2017). https://energy.gov/eere/femp/institutional-change-federal-energy-management.

3.5   The Strategic Energy Effectiveness Framework (SEE)

These theories have been important in the development of the Framework, which borrows from them a number of different principles and methodologies:

- Enterprises are complex systems; therefore energy and water management should be strategic.
- Excellence is far better than mediocrity.
- Lines of authority and control can only be effective if they are well designed.
- Energy management is a skill that can be taught, not an arcane technical discipline.
- Good design improves productivity.
- No management theory works if people are left out of the picture, and Groupthink can drive organizations in both positive and negative ways.
- Quality is key, even if short-term profits suffer – as long as the goals or objectives can be met.
- Feedback loops enable both improvement and persistence.
- Negative impacts on the environment should be reduced.
- Measurement and verification methodologies can vary dramatically, so they should be agreed upon in advance.

3.6   Data and Information

One further point regarding management theories and history is important, in a time when big data surrounds us. From the discipline of financial analysis where financial presentations are carefully crafted to choose the most appropriate and relevant data and present it in a well-crafted analysis, managers know that data in and of itself is useless. That insight is no less true for energy programs and projects, which should be connected to the value and purpose of the institution. Or, as noted in Frank Diana's blog on Business Analytics under "Business Analytics will Enable Outcomes": "Business Analytics focuses on answering questions such as why is this happening, what if these trends continue, what will happen next (predict), what is the best that can happen (optimize). There is a growing view that prescribing outcomes is the ultimate role of analytics; that is, identifying those actions that deliver the right business outcomes. Organizations should first define the insights needed to meet business objectives and then identify data that provides that insight. Too often, companies start with data."[6]

---

[6] Diana. (2011). https://frankdiana.net/2011/03/19/the-evovling-role-of-business-analytics/.

References

Sources Used in the Chapter

(n.d.). Retrieved from file:///C:/Users/Sandra/Downloads/RWP15_004_Stavins.pdf.

50001, I. (n.d.). ISO 50001. Retrieved from Energy.gov: https://energy.gov/ISO50001. Accessed 3/27/2017.

Bosman Readmore, M. (2009). The historical evolution of management theory from 1900 to Present: The changing role of leaders in Organizations. http://faculty.wwu.edu/dunnc3/rprnts. historyofmanagementthought.pdf.

Carayannis, E., et al. (2014). *Business model innovation as lever of organizational Sustainability*. New York: Springer Science+Business Media. https://doi.org/10.1007/s10961-013-9330-y.

Efficiency Valuation Organization. (2009). http://mnv.lbl.gov/keyMnVDocs/ipmvp. Retrieved from evo-world.org: evo-world.org/en/.

Grodsky, T. (n.d.). History of management thought. Retrieved from Faculty.wwu.edu: http://faculty.wwu.edu/dunnc3/rprnts.historyofmanagementthought.pdf.

SustainAbility. (2014). 20 Business Model Innovations for Sustainability. SustainAbility.

Young, R., et al. (2014). *The 2014 international energy efficiency scorecard report E1402*. Washington DC: American Council for an Energy-Efficient Economy.

Additional Links for Those Interested in Learning More

History of management theory.: http://www.strategicleadershipinstitute.net/news/the-historical-evolution-of-management-theory-from-1900-to-present-the-changing-role-of-leaders-in-organizations-/.

Contemporary management theory.: http://managementhelp.org/management/theories.htm. Accessed 4/29/2016.

http://www.saylor.org/site/wp-content/uploads/2013/02/BUS208-2.1-Historical-and-ContemporaryTheories-of-Management-FINAL.pdf. Accessed 4/29/2016.

Excellent general summary resource.: http://www.nwlink.com/~donclark/history_management/management.html. Accessed 4/29/2016.

https://www.lean.org/WhatsLean/History.cfm. Accessed 3/27/2017.

Harvard Business Review., https://hbr.org/2014/07/managements-three-eras-a-brief-history/

Peter Drucker – Management by Objectives. http://communicationtheory.org/management-by-objectives-drucker/. Accessed 4/29 and Managing by Results http://www.yourarticlelibrary.com/business-management/6-major-contributions-of-peter-drucker-to-management/27900/. Accessed 4/29/16.

Japanese Quality Control/Deming. http://www.wtec.org/loyola/ep/c6s1.htm. Accessed 4/29.

Quality assurance. http://www.quality-assurance-solutions.com/basic-tools-for-process-improvement.html. Accessed 4/29.

Process improvement -- for commanding officers. https://balancedscorecard.org/Resources/Articles-White-Papers/Process-Improvement-Tools. Accessed 4/29.

Six Sigma process improvement. https://en.wikipedia.org/wiki/Six_Sigma. Accessed 4/29/16.

ISO 9000. http://www.sis.pitt.edu/mbsclass/standards/martincic/isohistr.htm. Accessed 4/29/16.

ISO 50001 in 2011, continuous improvement http://www.sis.pitt.edu/mbsclass/standards/martincic/isohistr.htm. Accessed 4/29 and also sublink to ISO 50001 2011; in addition http://energy.gov/eere/amo/iso-50001-frequently-asked-questions 4/29.

IPMVP.   http://www.coned.com/energyefficiency/PDF/IPMVP%20Vol%201_2010_En.pdf. Accessed 4/29.

http://www.leonardo-energy.org/resources/190 intro to ISO50001 Energy management, application note can be downloaded, 7/17/2017.

Chapter 4
The History of Energy Use

4.1 Introductory Comments

Now that readers have an appreciation for the important role energy and water play in the world today and an idea that enterprise-oriented analyses can help understand and address utility uses and issues, this chapter focuses more closely on the nature of energy itself and the history of its use. A practical approach to energy projects relies on a basic understanding of the history of energy use in organizations, which this chapter provides.

4.2 Energy and Water in the Wild

Energy and water exist in the wild of course; they are part of the fabric of this planet, and without them life could not exist. In today's world, however, they are transformed to be useful at the scale needed for economic activity and other purposes. Originally, both energy and water were used close to where they were found or produced; many towns began life around a stream with a mill, where both water and power were available.

Energy must be transformed from its raw state as stored water, oil, natural gas, solar rays, wind, and other sources to accomplish the many tasks now required of it; over the past 150 years, generating energy from fossil fuels has been the predominant methodology for accomplishing that transformation. The electricity, natural gas, and other fuels are then delivered to the location where they will be used. The same is true of water, which is mined and treated to become potable and then treated again once it becomes wastewater.

Of course, distributing water from where it is found to where it is required has a very long history, going back to the aqueducts of the Romans or the Qanats of

© Springer International Publishing AG, part of Springer Nature 2018
S. McCardell, *Energy Effectiveness*, https://doi.org/10.1007/978-3-319-90255-5_4

Afghanistan and the Gobi desert, which brought water from mountains to population centers thousands of years ago.[1]

Electrical distribution started with the taming of electricity through discoveries made by inventors such as George Westinghouse, Thomas Edison, and Nikola Tesla[2]; other forms of energy (such as natural gas and petroleum) spread with increased transportation and distribution options. The human history of energy use has been one of adaptation.

- Early humans, once they discovered the advantages of fire, kept coals close at hand and brought wood to their camps for warmth, light, and cooking.
- Nepali villagers used to thresh and winnow their rice by hand; village roads brought passing vehicles to thresh the grain.
- Natural gas was used or flared near where it was sourced until large networks of natural gas pipelines, and other distribution mechanisms began to allow Venezuelan gas to be sold in New York City.
- Petroleum products are distributed worldwide – providing energy as well as the ability to transport energy in diesel-fueled trucks or ships. Petroleum distillates can also be found in foods, containers, buildings, and thousands of other products.

In Europe, the United States, and many other countries, all organizations depend on energy in its different forms. Countries where energy is less reliable aspire to reach that same point where their institutions and citizens can depend upon reliable, good quality energy.

4.3   The Industrial Revolution

Although the Industrial Revolution began before the widespread availability of electricity, the changes since 1882 when Thomas Edison designed and built the Pearl Street Generating Station have been increasingly rapid. The Pearl Street Station provided direct current to local customers who were then able to purchase incandescent lamps, which Edison had invented in 1879. Power was supplied by six dynamos, reciprocating steam engines drawing from four coal-fired boilers. By the end of the nineteenth century, alternating current systems based on inventions from Nikola Tesla and many others, promoted by visionary industrialists such as George Westinghouse, developed into the larger and more extensive versions of the electrical distribution networks we have today. Energy generation today relies predominantly on fossil fuel sources extracted and used in ways that have not changed significantly over the last 150 years, as can be seen in the photo of Valmont Station near Boulder, built in 1921 (Fig. 4.1).[3] Crews unload pulverized coal at a site which

---

[1] Macpherson. (2017). 7/16/2017.

[2] King. (2011). Accessed 7/16/2017.

[3] Sulzberger. (n.d.). Accessed 16 Sept 2016.

Fig. 4.1 Loading coal at Valmont Station. (Electric Light & Power (2013). Accessed Sept 16, 2016)

now houses a 229 MW power plant that generates electricity from low-sulfur coal and natural gas.

Once electricity could be generated and distributed, long-distance transmission and distribution lines were installed, and in densely populated areas, both individuals and institutions came to rely on the services provided. Creative inventors and businesses began to develop products to take advantage of that availability – refrigerators, plastic molding machines, televisions, computer systems, brightly lit signage, mobile phones, etc. The list is endless. And although the benefits of these changes and products are great, progress has not always been positive.

4.4   The Idea of Using Less Energy

Energy is generated, transmitted, and used, an inherently inefficient process with large losses (or waste) along the way. With what is known as a paradigm shift, learning to see something in a completely new way, that waste can become a resource. Using such an approach, energy efficiency and energy conservation programs can be seen as sources of energy or negawatts, in the terminology developed by Amory Lovins of the Rocky Mountain Institute.[4] By recapturing electricity which has

---

[4] Lovins (1989).

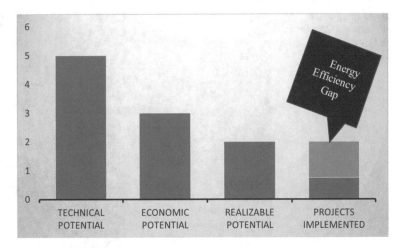

Fig. 4.2 Energy efficiency gap

already been paid for but would otherwise be lost instead of used, an enterprise essentially "produces" more energy for its own use at negative or very low cost.

Efforts to increase the efficiency of all operations inside an enterprise have been ongoing since Taylor attempted to make mine workers more efficient or the TQM (total quality management) movement tried to reduce the cost of inefficient production and poor quality. In highly energy-intensive industries, reducing energy use and cost has often been part of increasing manufacturing efficiency, but the potential for savings remains and increases with new technologies and methods.

There are literally thousands of individual ways to reduce utility cost and use, but few of them are applicable everywhere. The industry distinguishes between those project proposals which:

- Are leading edge or bleeding edge, appropriate for technically advanced institutions interested in serving as test cases, who can troubleshoot issues and report lessons learned
- Have technical potential, those which are technically feasible, without considering other aspects of the project
- Have economic potential or are economically viable without considering other parameters
- Have realizable potential after including political, financial, or other implementation barriers

A small subset of those projects with realistic potential are actually implemented; the delta between those implemented and those not implemented is called the energy efficiency gap (Fig. 4.2).

Energy, when used, has two elements: the energy itself (watts, therms, cubic feet, etc.) and the time during which it is used. This is true of electricity, natural gas, water, and other types of energy which are not stored on-site but delivered on demand. In other words, the amount of natural gas used by a furnace is a function of the amount of gas required to operate the furnace, multiplied by the amount of time the furnace runs and draws gas through the pipeline to the plant. Both the instantaneous amount of energy and the time for which it is required can be controlled by using different techniques. Early efforts to reduce energy use focused on energy conservation or turning equipment off when it was not needed. Some readers may remember what came to be known as US President Jimmy Carter's Sweater Speech of 1977, which recommended lowering thermostats as a response to the oil embargo.[5] It was an appropriate, if highly unpopular, strategy.

With new fears of high prices and low reliability for petroleum resources, manufacturers started designing more efficient products that required little or no human intervention or awareness. Those efforts came to be known as energy efficiency programs.

More recently, the realization that people engaged in practicing energy conservation and machines or equipment designed to be energy efficient, can work together to dramatically improve results has begun to spread. Deep energy effectiveness with savings on the order of 50% requires a change in behavior and culture as well as an integrated, systemic approach. The process is iterative, with many small yet consequential steps to take along that path.

As energy effectiveness increases, enterprises can become stronger and more competitive. And as this occurs, the economic benefit of wasting fewer fossil and renewable energy resources has been significant. In the United States, for example, energy efficiency improvements are responsible for 60–75% of the increase in energy productivity from 1970 to 2013, a step toward maintaining American competitiveness in the world.[6] In addition, with the energy component of products and services included in the GDP being half what it had been 40 years previously, aggregate production costs for goods and services were lower.

US economic output expanded more than 300% over the past 40+ years, and energy use grew only 50%.[7] Energy effectiveness played a significant role, as did the idea of on-site generation and the transition to less energy-intensive industries. Energy effectiveness also mitigated the effect of increasing average home size, vehicle miles, and proliferation of energy-consuming devices at work and at home.

Generated energy and water are used principally in buildings, the industrial sector, and transportation, the last of which is not covered in this book. There is

---

[5] https://www.youtube.com/watch?v=MmlcLNA8Zhc. Accessed 3/27/2017.

[6] (Alliance to Save Energy, 2013) page 3.

[7] (Alliance to Save Energy, 2013) page 4.

a close relationship between energy and water, sometimes referred to as the energy-water nexus because approximately 30% of total water use is for producing energy and approximately 30% of energy use is for treating and delivering water.

4.5   Energy Effectiveness Drivers

Management systems, measurement/monitoring systems, and new technologies are now at the point where energy effectiveness is increasingly achievable. Growing interest in energy effectiveness is a result of the following trends, among others:

- Volatile energy markets, beginning with the 1973 oil embargo, which demonstrate the risks of relying on energy sources controlled by others – whoever those others might be
- Volatile energy prices, which impact organizations' ability to plan, budget, and control costs
- Increasing prevalence of energy efficiency as a design goal, resulting in better and more efficient products
- Government policies which provide incentives for energy conservation and energy efficiency
- Utility company realization that negawatts sometimes provide the least expensive way to meet their obligations
- Public concern about generation methodologies which increase greenhouse gas emissions and therefore global warming
- Competitive cost pressures which cause management to focus on cost reduction where possible
- Increased understanding of decision-making processes and educational approaches to increasing energy effectiveness
- Increased competition worldwide, especially from low-wage countries
- Reduced costs and increased reliability for alternative energy products and systems
- Water scarcity
- Internet of Things (IoT) enabling localized, cost-effective data collection
- Increases in utility market and reliability risk
- Reduction in technology risk because most technologies are well proven
- Increasing variety of energy project financing options for all types of enterprises
- Increased availability of and access to on-line and other tools to analyze technical options with financial metrics

The ACEEE International Scorecard grades each country on their energy efficiency for the year. As the map below demonstrates, the interest and investment in energy efficiency programs are worldwide, with Germany ranking highest in 2016 in buildings and industry as it has for many years (Fig. 4.3).

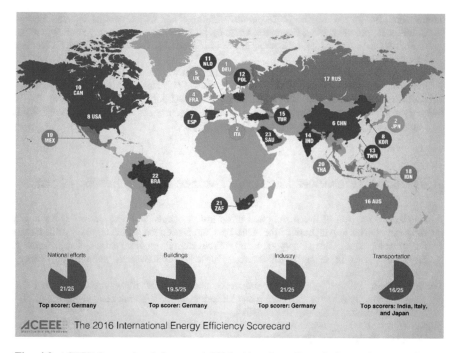

Fig. 4.3 ACEEE International Scorecard 2016. (American Council for an Energy Efficient Economy (2016). Accessed 7/17/2017)

References

References Cited in the Chapter

Alliance to Save Energy. (2013). *The history of energy productivity*. Washington, DC: Alliance to Save Energy. Retrieved from http://www.ase.org/sites/ase.org/files/resources/Media%20 browser/ee_commission_history_report_2-1-13.pdf.

American Council for an Energy Efficient Economy. (2015, February). aceee.org/blog. Retrieved from ACEEE: http://aceee.org/blog/2015/02/ why-we-don%E2%80%99t-have-choose-between-ener.

American Council for an Energy Efficient Economy. (2016). World Energy Efficiency Scoreboard 2016. Retrieved from Aceee.org: http://aceee.org/sites/default/files/image/topics/2016-world-scores.png.

Carter, J. (1977, February 2). Miller Center for Public Affairs. Retrieved from Youtube: https:// www.youtube.com/watch?v=MmlcLNA8Zhc.

Electric Light & Power. (2013). www.elp.com. Retrieved from Electric Light and Power: http:// www.elp.com/articles/slideshow/2013/08/a-look-back-at-electric-utility-history/pg003.html.

King, G. (2011, October). Smithsonian.com. Retrieved from www.smithsonianmag.com: http:// www.smithsonianmag.com/history/edison-vs-westinghouse-a-shocking-rivalry-102146036/. Accessed 7/16/2017.

Lovins, A. (1989). *The Negawatt Revolution - Solving the CO2 Problem* . Green Energy Conference, Montreal 1989, Keynote Address (p. http://www.ccnr.org/amory.html). Montreal: Canadian Coalition for Nuclear Responsibility. Retrieved from Canadian Coalition for Nuclear Responsibility.

Macpherson, G., et al. (2017). Viability of karezes (ancient water supply systems in Afghanistan) in a changing world. *Applied Water Science, 7*, 1689–1710, 1690. Retrieved from https://link.springer.com/content/pdf/10.1007%2Fs13201-015-0336-5.pdf.

Sulzberger, C. (n.d.). Milestones. Retrieved from Engineering and Technology History Wiki: http://ethw.org/Milestones:Pearl_Street_Station,_1882.

Additional References for Those Interested in Learning More

A complete description of this history and those trends is outside the scope of this book, but for readers who are interested, the Alliance Commission on National Energy Efficiency Policy publication "The History of Energy Productivity" provides an excellent summary. https://www.ase.org/sites/ase.org/files/resources/Media%20browser/ee_commission_history_report_2-1-13.pdf.

For a fascinating series of photographs and stories about electric utilities dating back to 1922, please see http://www.elp.com/articles/slideshow/2013/08/a-look-back-at-electric-utility-history/pg001.html. Accessed 16 Sept 2016.

https://www.worldenergy.org/wp-content/uploads/2014/03/World-Energy-Perspectives-Energy-Efficiency-Technologies-Overview-report.pdf.

Chapter 5
Energy Flows and Water Flows: Types of Energy

5.1 Energy Flows

The Sankey diagram shown in Fig. 5.1, produced by Lawrence Livermore National Labs for the US Department of Energy, shows estimated sources and uses of energy in the United States in 2015. Electricity is generated from solar, nuclear, hydro, wind, geothermal, natural gas, coal, biomass, and petroleum resources – with coal and natural gas together representing about 2/3 of production. Of most interest in the context of this book, however, is that 38.0 quads of energy (A quad is equal to 1 quadrillion Btus or British thermal units) are used to generate 12.6 quads, with 26.4 quads – a third of the total – being wasted. The gray bands show the wasted/unused resource as rejected energy; this is the resource more effective energy-related practices and methodologies can recapture.

Rejected energy is also a factor in the residential, commercial, industrial, and transportation sectors, with the result that energy services are 38.4 quads and rejected energy for that sector 59.1 quads. Energy services and productive energy are different terms for what energy enables – light for homes, power for manufacturing equipment, transportation delivering products to market, and employee comfort. In aggregate, 40% of the energy produced provides those services; the other 60% is generated and paid for – but thrown out along the way.

As a comparison, the Sankey diagram in Fig. 5.2 is for the European Union, with data from 2006. It shows a similar pattern in terms of generation and what are here termed energy losses of 62%. (The somewhat lower efficiency for the European Union may be explained in part by timing differences and variations in energy sources.)

As these charts demonstrate, current economic activity relies on extracting fossil fuel resources deposited over the course of millions of years, millions of years ago, in what can be seen as a geologic blink of an eye. The extracted resources have then been used for power, warmth, and transportation in a span of less than two centuries. Over 60% of the energy generated becomes waste, and most of that waste is heat which escapes either inside a facility or into the environment.

© Springer International Publishing AG, part of Springer Nature 2018
S. McCardell, *Energy Effectiveness*, https://doi.org/10.1007/978-3-319-90255-5_5

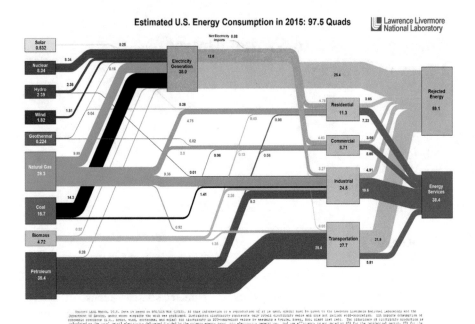

Fig. 5.1 Sankey diagram, US energy consumption. (Lawrence Livermore National Laboratory (2016). Accessed 9/16/2016)

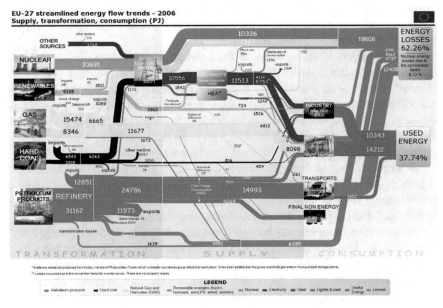

Fig. 5.2 Sankey diagram, European energy consumption. (Sankey diagram for European (EU27) energy flows from (European Union). Accessed 8/4/2017)

5.2 Types of Energy

Energy, of course, exists in myriad forms; it helped form the universe, and there are several types. According to the first law of thermodynamics, in a closed system, energy and matter cannot be created or destroyed, because the mass of the system must remain constant. Inside such a system, matter and energy can both be changed from one state to another.

Energy is briefly described as "the capacity of a physical system to perform work." A more scientific definition is "Energy is an indirectly observed physical quantity which is the ability of a body or a system to do work."[1]

Kinetic energy is generated by something that is moving, while potential energy is stored. Within those categories, energy takes many forms, some of which are more useful for the work we need than others. The table below lists some of the most intriguing or most useful forms of energy, along with some of the work those types of energy can do (Table 5.1).

Table 5.1 Types of energy and potential uses

Type of energy	Description	Potential uses
Heat/Thermal	Heat is either directly or indirectly from the sun; it is a by-product. Energy is produced by moving particles	Heating buildings or water, cooking, cooling (through a heat exchanger or ice-cream maker)
Chemical energy	Transition of fuels stored chemically	Fire, hydrolysis, battery
Magnetic energy	Energy which causes a "push" or a "pull"	Motors use magnetic energy. "Maglev" transportation is under development
Mechanical	Energy produced by objects in motion	Moving vehicles, animals, etc.
Electrical	Energy from particles moving through a wire or to another destination	Thousands of uses, transmitting through wires, lightning/arcs
Electromagnetic	Combined energy sources	Visible light, infrared radiation, radio waves
Solar	Nuclear fusion in the sun	Directly or indirectly provides power and heat
Elastic	Energy stored in objects that are stretched	Children's toys, process equipment
Gravitational	Energy stored in an object above the planet's surface	Pendulums, building structures
Nuclear	Energy which is stored in the center (nucleus) of particles. Fission and fusion are two types	Storage for later use
Chemical	Energy which is stored in food or fuel	Gasoline, energy for marathons, batteries

---

[1] Admin. (2012 updated 2013). Accessed 3/23/2017.

Energy can be converted from one form to another, which is how electricity is produced. In that process as in every conversion process, energy is wasted as heat; therefore utilities and engineers reduce the number of conversions whenever possible. To produce electricity, coal and gas are burned in a generation facility, powering a turbine to generate electricity that is then transmitted across hundreds of miles to the end user.

Most energy for buildings and processes is provided through electricity and natural gas. Natural gas occurs naturally underground, sometimes in association with liquid petroleum, and is refined to a standardized composition before being transported or sold. It can then be used or transformed into other forms of energy. Other types of gas are also available for the same purposes, including methane from coal beds, landfill gas, synthetic gas, and biogas.

Alternative energy systems situated on-site can be of benefit for many different reasons including increasing reliability, reducing cost, reducing waste, and enhancing architecture, although they must be carefully chosen and designed to be appropriate in the local environment and for the intended purpose.

5.3   Sources of Energy

Electricity exists in its own form only for a moment, as in a lightning strike. Even knowing that, however, it is difficult to understand that electricity must be used immediately when generated – it can only be stored if converted to another form of energy. According to the US Energy Information Administration, 2015, electricity generation totaled about 4 trillion kWh (or kilowatt-hours), 67% of which came from fossil fuels. The full list of major energy sources and percent share of total US electricity generation in 2015 shows:

- Coal = 33%
- Natural gas = 33%
- Nuclear = 20%
- Hydropower = 6%
- Other renewables = 7%

  - Biomass = 1.6%
  - Geothermal = 0.4%
  - Solar = 0.6%
  - Wind = 4.7%

- Petroleum = 1%
- Other gases = <1%

(Preliminary data based on generation by utility-scale facilities)[2].

---

[2] US Energy Information Agency. (n.d.). Accessed 3/24/2017.

5.4 Sources and Carriers

Energy can also be segregated into energy sources and energy carriers. Natural gas is a source, while electricity is a carrier; both fill different roles in the energy-supply chain from primary sources to the end user. As the energy industry evolves from the current heavy reliance on fossil, both energy sources and energy carriers will evolve. Hydrogen technologies and smaller-scale decentralized energy generation systems, for example, may become more important.

As previously noted, some energy is inevitably lost when converted from one type to another; it is much for efficient therefore to use source energy whenever possible. Carriers are required in many situations, however, particularly where there are technology and transportation issues. Where alternative energy sources are available nearby or can be situated on-site, that can be an attractive option because then the primary source is used near where it is produced. Without the necessity for conversion to a carrier, there are attendant savings in capital investment, energy losses, and carbon emissions. It is a complicated picture, as shown in Fig. 5.3.

The full menu of energy options is extensive, but if a manager starts with the energy available and the energy needs of one particular enterprise or location, the number of options narrows significantly; generally there are at most two or three fossil fuel options and similar numbers of alternative sources. Biomass is not available in the desert, for example, nor are wind turbines appropriate in most towns.

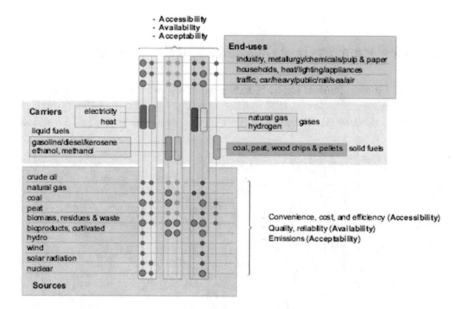

Fig. 5.3 Energy sources, carriers, and end uses. (Intergovernmental Panel on Climate Change (n.d.). Accessed 8/4/2017)

5.5   Water Flows

Water resources are important to organizations and for economic development as well as to the communities and people with whom the organizations are interrelated. In fact, if current trends continue, water may become an even more critical resource and potential risk than energy. Prices for water are generally unsustainably low, but aging infrastructure and resource depletion would appear to presage challenges and risks in the near future. With only 0.5% of the water on earth available to meet all freshwater needs as shown in Fig. 5.4, and with water sources being depleted in many heavily populated areas of the world as shown in Fig. 5.5, the need to conserve water resources can only increase.

According to the same WCSB report, 22% of water use worldwide is for industrial purposes. In low- and middle-income countries, the average is 10%, and in high-income countries, it is 59%. Those statistics combined with the picture of water resources deficits above highlight the critical nature of water resources for organizations and managers.

Water is becoming more and more scarce, aquifers are becoming depleted, infrastructure improvements are needed, and yet the price paid for potable water remains low in much of the developed world. The situation is beginning to be addressed in pockets, where local governments provide incentives to reduce water use. Reducing water use, and reusing it where appropriate, are imperative for most managers; the results are reduced costs and usage of both water and energy in an organizational context.

5.6   How Humans Have Generated and Used Energy

Over the centuries, humans have accomplished much work using different types of energy from multiple sources, with increasing reliance on what that energy enables. Fossil fuels such as petroleum, coal, and gas permeate all activities, permitting

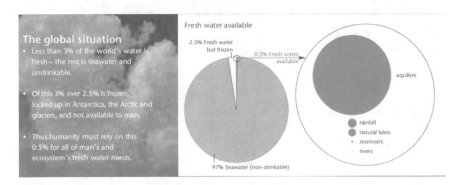

Fig. 5.4 The global water situation. (Fry (2005/Reprint 2006) page 1)

Annual renewable water (m³/person/year)⁵

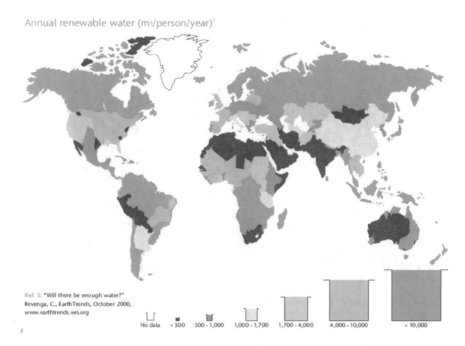

Ref. 5: "Will there be enough water?"
Revenga, C., EarthTrends, October 2000,
www.earthtrends.wri.org

No data   < 500   500 - 1,000   1,000 - 1,700   1,700 - 4,000   4,000 - 10,000   > 10,000

Fig. 5.5   Annual renewable water. (Fry (2005/Reprint 2006) page 2)

humans to live in ways impossible previously. The consequences of this reliance on fossil fuels are now being felt, however; water, soil, air, and living beings are saddled with the consequences of releasing an immense storehouse of fossil energy in a relatively short period of time.

Direct costs for alternative energy systems are plummeting, and the costs borne by society at large for pollution and unsustainable fossil fuel-based generation are increasing. This is an inflection point, where fossil fuel-based energy production is likely for the foreseeable future, but renewable energy technologies are either already at parity (of the same cost as fossil fuels) or projected soon to reach that point. Economically, alternative energies are becoming a more sensible choice.

Perhaps humans first learned to conserve energy by wearing skins to hold body warmth or by living in caves to stay cool during hot days. Fire allowed humans to change the chemical energy of wood into thermal energy and cook food. Tar and pitch were used for torches, and later whale blubber was used for light. Water wheels ground corn and pack animals provided transportation.

Interestingly, fossil fuels are sometimes referred to as traditional sources of energy, although this particular tradition is less than 200 years old while alternative energy sources date back thousands or tens of thousands of years (Fig. 5.6).

Windmills were evidently used in both Persia and China around 2000 years ago, primarily for water pumping and grinding grain[3]. The first uses for wind for

---

[3] Telosnet. (n.d.). Accessed 4/29/2016.

Fig. 5.6 Stone windmill. (Photo via <a href = "https://visualhunt.com/re/b8407e"> VisualHunt</ a> commercial royalty free photo https://visualhunt.com/photo/196446/)

transportation date back even further, to the first sailboats some 5000 years ago in Mesopotamia[4] and at about the same time in Polynesia[5] (Fig. 5.7).

Sailing vessels and wind turbines still operate today in different forms, and the principles by which they operate are being adapted in many other ways – from undersea turbines in bewildering variety, among them underwater kites,[6] to turbines used for generating power inside pipes (Fig. 5.8).

The history of electricity is a story of multiple realizations, discoveries, and inventions that allowed humans to generate and more effectively use a force that had been known for thousands of years[7,8]. In that process, one significant milestone was when George Westinghouse won the contract to build the first generators at Niagara Falls in 1893, a step toward the development of today's electrical distribution grid (Fig. 5.9).

A strong competitor to Westinghouse and leading light of the time was the prolific inventor Thomas Edison, best known for developing the incandescent light bulb – a design which has lasted through to the present day. More importantly in the context of this book, he also developed utility and transmission systems to power the light bulbs and other appliances he had invented (Fig. 5.10).

---

[4] Merlin. (2009). Accessed 4/29.

[5] Wikipedia. (n.d.). Accessed 4/.

[6] Merchant. (2015). Accessed 4/29.

[7] Columbia Electronic Encyclopedia. (2012). Accessed 4/29/2017.

[8] Henry. (2017). Accessed 4/29/2017.

Fig. 5.7 Trading ship. (Photo credit: <a href = "https://www.flickr.com/photos/library_of_congress/3174194245/" > The Library of Congress</a> via <a href = "https://visualhunt.com/re/a96d87">Visualhunt</a> / <a href = "http://flickr.com/commons/usage/"> No known copyright restrictions</a>)

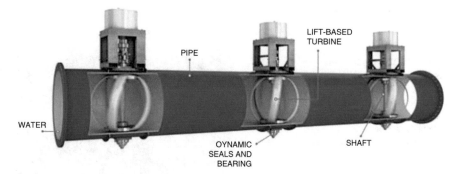

Fig. 5.8 Cross section of pipe being tested in Portland by Lucid. (Valentine (2015). Accessed 4/29/2016)

Although transportation fuels are not the subject of this book, a few quick comments are relevant because diesel fuel and natural gas are used for transportation fuels as well as for energy inside facilities. In addition, transportation fuels can be converted to energy carriers and then used by organizations for other activities; the same source fuels are just used in different ways.

During the Industrial Revolution, coal was used to power steam locomotives and steam power plants[9]. It was plentiful in many local areas, easy to access and to burn.

---

[9] Mitsubishi Heavy Industries. (n.d.). Accessed 4/29/2016.

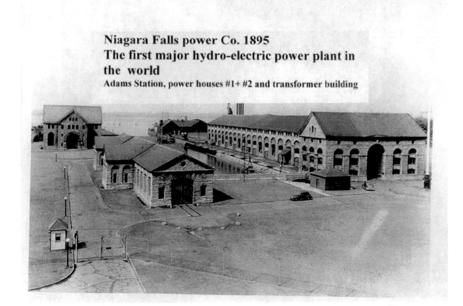

Fig. 5.9 Niagara Falls Power Company 1895. (Dr. Vujovic, President Tesla Memorial Society of New Nork (n.d.). Accessed 3/24/2017)

Fig. 5.10 Thomas Edison and his incandescent light bulb. (Quotes (2016). http://www.edisonmuckers. org/thomas-edison-inventions/. Accessed 3/24/2017)

Fig. 5.11 A vintage diesel
engine. (Ury (2013).
Accessed 4/29/2017)

In fact, coal is still found lying on the ground to be picked up and burned in some
places, and there may be readers who remember room-sized oil furnaces for home
use which started life as coal furnaces. Petroleum was at first used only for kerosene
(still a staple for cooking and light in many parts of the world), but the industry soon
unleashed a flood of innovation which adapted or developed petroleum products to
power vehicles, planes, and trains; the waste from those processes turned into plas-
tics, cosmetics, and thousands of other products we use today[10] (Fig. 5.11).

In addition to steam, many other energy sources were used in the early years of
the automobile, with Rudolf Diesel's compression-ignition engine delivering 75%
efficiency in contrast to 12% from steam engines. The "diesel" was originally powered by peanut and other plant oils.

Solar power is the original energy source, leading to the creation of every other
fuel source over time. Solar power systems, on the other hand, have a shorter history
but a rapid pace of innovative activity as shown in the timeline in Fig. 5.12 The
question, "What will Tomorrow Bring?" is apt. (The timeline can be downloaded
from   http://www.theecoexperts.co.uk/sites/default/files/filemanager/The_Solar_
Timeline.jpg.).

---

[10] McLamb. (2010). Accessed 4/29/2017 http://www.ecology.com/2010/09/15/secret-world-energy/.

Fig. 5.12 The solar
timeline. (The Eco Experts
(2013). Accessed
4/29/2017)

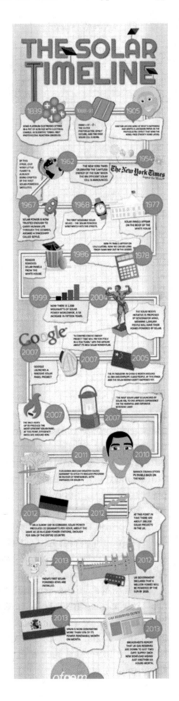

5.7   Alternative Energy Types Relevant for Enterprises

There are ways to buy the attributes of renewable energy without actually buying that energy. Airplane travelers, for example, can carbon mitigate their jet fuel by purchasing carbon offsets from organizations such as the Carbon Fund, www.carbonfund.org. Renewable Energy Certificates (RECs) or Green tags are similar products purchased by individuals or businesses to fund alternative energy projects to which they have no other connection, at an extra cost. Those types of alternative energy products can have great impact and may support the organization's mission, but they are outside the operations of an enterprise and not addressed in this book which examines only those projects which would generate power for the facility.

Appropriate alternative energy projects to be considered for any specific facility are only those which are or can be available locally. In practice, that constrains the choice to the following:

- Passive solar design and building techniques
- Solar photovoltaic electricity
- Solar thermal heat or cooling
- Wind generation
- Low-temperature geothermal (ground source heat pumps)
- Biomass
- Waste heat recapture
- Energy storage systems for demand cost reduction

The other key consideration is that management evaluate only those systems which address relevant issues and uses. Each type of system has the potential to accomplish different things as shown in Table 5.2.

Although solar power does have a long history, in recent years significant advances in solar panel efficiency, storage capacity, and manufacturing techniques have led to dramatic decreases in the cost of solar power systems. According to the Solar Energy Industry Association, the price of a solar installation dropped more than 60% from 2006 to 2016 and that year solar installations in the United States represented the highest percentage of electrical capacity added to the grid for the first time at 39%; natural gas was 29%, and wind was 26% as shown in Fig. 5.13.

Solar PV systems are not applicable everywhere, but they are flexible and can be adapted to multiple environments and facilities. Other sources of power can be challenging to install at commercial or industrial facilities in a constrained space or in some environments but merit at least initial consideration.

- *Wind*: Wind turbines come in many different varieties. The large turbines used for utility scale power are often located at a distance from towns or cities and require large transmission lines. Smaller wind turbines can be located in an appropriate wind regime to provide power to a community or a facility, as long as they comply with zoning laws.
- *Biomass*: It is an extremely varied category consisting of biofuels, biopower, and bioproducts such as synthetic gas. Feedstocks are often waste products such as sewage, agricultural residue, cooking oils, manure, plant oils, etc. Biomass

Table 5.2 Alternative Energy Technologies and Potential Uses

Technology	Characteristics	Potential uses	Examples
Passive solar design and building	Design and siting to take advantage of wind, sun, light	Daylighting, passive cooling, earth sheltering, use of thermal mass	Light shelves to direct daylight, southern windows for heat in N. latitudes
Solar photovoltaic electricity	Solar panels generate power. Can be tied to the electrical grid, stand alone, or hybrid	Different configurations have different uses including risk reduction and cost savings	Car ports covered with PV panels, PV systems for backup power
Solar thermal heating or cooling	Solar thermal systems heat air or liquids. With heat exchanger, it can also cool	Water heating or preheating, air cooling, building heating	Heating water for pools or laundries
Wind generation	Wind turbines are either horizontal or vertical axis, require consistent wind speeds	Generate power, have very specific siting needs	Rural water pumps, lighting in far northern towns
Low-temperature geothermal/ GSHP	System of distributing the ground temperature to the building	Heating and cooling.	Schools, residences, etc.
Biomass	Many types and many different approaches. Varied feedstocks	Multiple potential uses from electricity generation to fuel and gas production	Space heating, electricity generation cooking

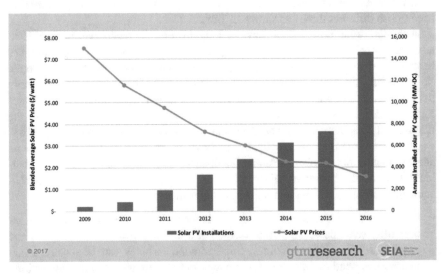

Fig. 5.13 Relationship between price and solar installations. (Solar Energy Industry Association (2017). Accessed 5/2/2017)

feedstocks and outputs are extremely varied and may be very attractive for organizations, but careful consideration should be given to the feedstocks and to the ways in which the output will be used.

- *Geothermal Systems/GSHP*: The term "geothermal system" is used to refer to what are essentially two different technologies: Small low-temperature geothermal systems or GSHP (ground source heat pumps) take advantage of the fact that ground temperature is reasonably consistent year-round. These use heat pumps to borrow the ground temperature to cool or heat a building. High-temperature geothermal systems are a special case appropriate only for power plants, which rely on geothermal activity deep underground to source steam used to generate electricity.
- *Combined Heat and Power*: For manufacturing operations or other facilities needing both power and heat, cogeneration (or combined heat and power, CHP) is worth considering. The fuel gas can be natural gas, landfill gas, methane, or another type of gas, and the waste heat from power generation is used for heating.
- *Waste Heat:* Waste heat recapture is an additional alternative energy resource because it is often cost-effective to recapture that heat for other uses. In any facility where both heating and cooling are required, waste heat from cooling can be used to preheat water or air. For example, in a commercial kitchen water can be preheated using waste heat from refrigeration.
- *Hybrid Systems:* Because alternative energy systems provide energy in highly targeted ways, they are often combined to increase overall production and reliability. The most common hybrid system is called a grid interconnect system, where a small wind system, for example, provides power when the wind is blowing and excess power is sent to the utility. When the wind is not blowing, the building draws power from the utility.

5.8   Energy Storage Systems

If alternative energy systems are used to generate power and the power will be used instantaneously, the systems can be simple. If, on the other hand, power is needed at a different time than when it is produced, it will need to be stored. Wind turbines often generate more power at night, but more power is needed during the day. A solar power system in England might produce little power in the winter, which is when that power would be most needed. Storage systems are thus a critical part of any renewable energy system, and there are several different approaches.

- *Grid interconnect:* The most common system in developed countries is more of a hybrid system than a storage system, but it is used for similar purposes. An installed system remains connected to the electrical grid, sending power to the utility when it is being produced, and drawing from the utility when the system is not producing what is required.

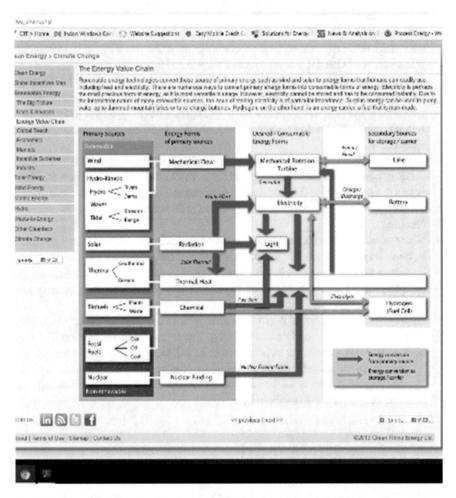

Fig. 5.14 Energy value chain. (Green Rhino Energy (n.d.). Accessed 5/6/2017)

- *Battery storage:* These systems are most common where there is no connection to the utility grid; they are sized to provide storage for anticipated needs over a number of days or hours.
- *Pumped water storage:* These systems are feasible where water can be pumped uphill to a lake when power is available and then delivered by gravity through a turbine which generates power when it is needed.
- *Other storage systems:* Ice storage, salt storage, flywheels, and other such approaches can also be used in particular circumstances, but their applicability is more restricted.

Figure 5.14 summarizes many of the points and technologies addressed in this chapter. As a manager, it is important to know that these technologies exist and can provide benefit. The specific choices and designs will be produced by experts, at which point it is the job of the manager to make the appropriate business choice between the opportunities presented. The remainder of this book focuses on organizations and the energy and water used inside them, with the goal of managing that energy and water.

References

Admin. (2012 updated 2013). Different types of energy sources. Retrieved from readanddigest.com: http://readanddigest.com/what-are-the-different-types-of-energy-sources.

Columbia Electronic Encyclopedia. (2012). Sandbox Networks, Inc. Retrieved from History of Electricity: http://www.infoplease.com/encyclopedia/science/electricity-history-electricity.html.

European Union. (n.d.). Google groups copied from EU. Retrieved from Energy Discussion: https://16625575041838918609.googlegroups.com/attach/f3692749112f9638/EU27_06.png?part=0.1&view=1&vt=ANaJVrFPHPNTdzkcLtEHOODI-aQfN0I9-tV0KlZI7GwmgTY-AAdxqZgdjhEbEkdxyCndXuF4RzKvR3NOxZ1eJunVuE0DCAtxY8Cvm3pyb31recP3Z-kY23zQ.

For those Who Are Interested, the Following Interactive Map of Energy Used in the United States might Be Intriguing: https://www.fastcompany.com/3062630/visualizing/this-very-very-detailed-chart-shows-how-all-the-energy-in-the-us-is-used%20with%20interactive%20version%20at%20http:/energyliteracy.com.

Fry, A., et al. (2005/Reprint 2006). Facts and Trends - Water. London: World Business Council for Sustainable Development, wbcsd.org.

Intergovernmental Panel on Climate Change. (n.d.). Publications and data 4.3.4 Energy Carriers. Retrieved from www.ipcc.ch: https://www.ipcc.ch/publications_and_data/ar4/wg3/en/ch4s4–3-4.html.

Lawrence Livermore National Laboratory. (2016). Flowcharts. Retrieved from llnl.gov Lawrence Livermore National Library: https://flowcharts.llnl.gov/content/assets/images/energy/us/Energy_US_2015.png.

McLamb, E. (2010, September 15). The secret world of energy. Retrieved from Ecology.com: http://www.ecology.com/2010/09/15/secret-world-energy/.

Merchant, B. (2015, January 2). Japan is Building Underwater Kites to Harness the Power of Ocean Currents. Retrieved from Motherboard.vice.com: http://motherboard.vice.com/read/japan-is-building-underwater-kites-to-harness-the-ocean-current-for-power.

Merlin. (2009, April 20). The evolution of the Sailboat and its effect on culture. Retrieved from serendip.brynmawr.edu: http://serendip.brynmawr.edu/exchange/node/4193.

Quotes. (2016, May 21). Thomas Edison Quotes. Retrieved from Playslack.com: https://playslack.com/post-96869/thomas-edison-quotes.

Solar Energy Industry Association. (2017). Solar Industry Data. Retrieved from SEIA.org: http://www.seia.org/research-resources/solar-industry-data.

Valentine, K. (2015, August 13). It's not a pipe dream: Clean energy from water pipes comes to Portland. Retrieved from Thinkprogress.org: https://thinkprogress.org/its-not-a-pipe-dream-clean-energy-from-water-pipes-comes-to-portland-16240cc33412/.

Wikipedia. (n.d.). Polynesian navigation. Retrieved from Wikipedia.org: https://en.wikipedia.org/wiki/Polynesian_navigation.

Part II
Organizations and Energy

Chapter 6
Energy Flows and Management Practices in Organizations

6.1 Doing Without Energy

The preceding chapters of this book have provided an overview of energy production and energy effectiveness, as well as general histories of management theories and energy use. Section II moves the focus to an internal one.

Perhaps one of the best ways to highlight the importance of energy and water in organizations is to make a list of what could be accomplished in their absence. In most cases, that list is very short: there is no lighting, computers do not run, air circulation is nonexistent, telephone systems don't work, heating and cooling systems are inactive, and industrial processes are completely shut down.

Organizations operating in locations prone to blackouts or brownouts and those which are designated as "critical" are required to install backup systems in preparation for the times when utilities are not functioning, but enterprises in areas with a robust utility infrastructure have come to rely on having electricity, natural gas, propane, water, or other utility services available at all times without real concern.

6.2 What Does Energy Provide? (Ignorance Is *not* Bliss)

Whether utility services are reliable or not, enterprises and individuals depend on energy and water – but don't purchase either commodity for their inherent primary benefits (e.g., oil or gas). It is the indirect benefits that are of interest – light, heat, or growing plants. Furthermore, the decision to purchase those side benefits is usually made without knowing how the utility will charge for the commodity that provides them. Worse yet, the billing process is such that payment is made for utility services purchased at least one or two months previously without an accounting for what was received or an analysis of the accuracy of the consumption metrics. No wonder

© Springer International Publishing AG, part of Springer Nature 2018
S. McCardell, *Energy Effectiveness*, https://doi.org/10.1007/978-3-319-90255-5_6

neither energy nor water is effectively managed within most organizations in spite of the significant proportion of total cost they represent!

There are thus several difficulties with the way utilities are purchased and handled.

1. What is purchased is not the same as what is desired; it is the secondary benefits of utilities which are of interest, for example:

    (a) Completed customer orders – online ordering and tracking system, integrated with shipping and inventory
    (b) Staff comfort – appropriate heating and cooling in offices
    (c) Visibility/outreach – nighttime lighting on signs and entrances, parking lot lighting
    (d) Building security – cameras, automatic door locks, lighting near entrances, etc.
    (e) Efficient manufacturing – including electrical devices, heating, cooling, and compressed air systems
    (f) Water availability – pumps to deliver water where it is needed, when it is needed, at the appropriate temperature
    (g) Product refrigeration – keeping foods safe and fresh
    (h) Staff and client health – providing sufficient air circulation and filtering for a healthy environment

2. What is purchased is always more than what is used due to losses along the way, as discussed in Section I.
3. It is in most cases (without real-time data) impossible to correlate actual usage with the actual utility cost. Without that information, finance managers are put in the position of signing checks for utility services without question; most think about energy use when looking in the rearview mirror, comparing this month's consumption with last month's or the same period last year.

As a consequence, the accountability and discipline required in other parts of an organization are absent when it comes to utilities. In a time of fast-moving competition and increased pressure on profit margins, this approach is no longer sustainable, so the new technologies and methodologies which enable managers to understand and control utilities in organizations of all sizes are welcome. Utilities are used in each and every activity, and they usually vary with the main utility using processes and systems.

6.3   Energy Usage Patterns

Energy usage patterns vary widely across sectors, as shown in Fig. 6.1. Manufacturing concerns desiring to reduce utility use will focus on machines and processes, while offices will focus on HVAC, lighting, and plug loads as shown Fig. 6.1.

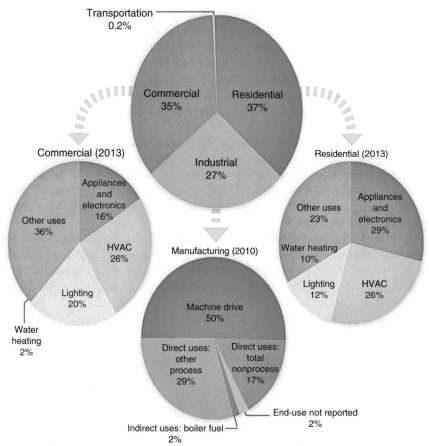

Fig. 6.1 Energy consumption by sector. (US Environmental Protection Agency (2017). https://www.epa.gov/energy/electricity-customers. Accessed 3/31/2017))

In summary, utility costs, often one of the most significant line items in any list of costs, are often ignored or seen as unmanageable – utilities are invisible. That invisibility serves not only to increase utility cost but other costs as well. These result from inadequate attention to the risks of aging infrastructure leading to predictions of higher risk and increased replacement costs in the future, weak utility-related cybersecurity procedures, and inadequate appreciation for the significant positive results that can be achieved by managing utility use.

6.4   Energy Flows Within Organizations

The now-familiar Sankey diagrams show manufacturing sector energy flows in two different categories. Process energy in Fig. 6.2 shows the energy which is used to produce products and which would normally be included in a cost of sales calculation (and therefore generate more interest than other utility uses). Fig. 6.2 shows that 4368 TBtu (trillion British thermal units) of energy or approximately 42% was lost in 2010. As manufacturing becomes more competitive globally, that waste needs to be addressed, as does the additional power required to cool buildings heated by process energy waste.

The same US Department of Energy report captures non-process energy in the manufacturing sector as shown in Fig. 6.3. Non-process energy use in the manufacturing context is very similar to energy use in other types of facilities, and therefore this diagram can also be used to understand energy flows in office buildings, shopping centers, and others. The total energy use is smaller, but a similar pattern prevails with 40% losses.

With the background of average energy flows in different types of facilities, the next step is to consider the energy flow inside the particular facility or organization of interest. The easiest approach is to install monitors which capture the information and present it appropriately, as does the Energy Flow Assessor from Enerit

Process Energy (TBtu), 2010

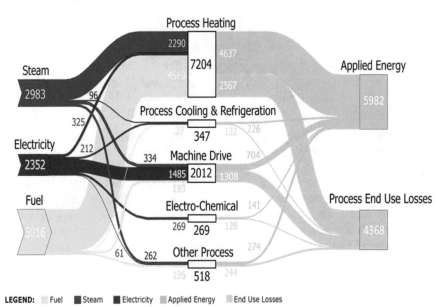

Fig. 6.2 Process energy in the US manufacturing sector. (Office of Energy Efficiency and Renewable Energy (n.d.-b). http://energy.gov/eere/amo/static-sankey-diagram-process-energy-us-manufacturing-sector. Accessed 8/8/2016)

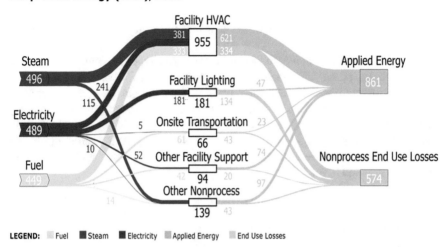

Fig. 6.3 Non-process energy in the US manufacturing sector. (Office of Energy Efficiency and Renewable Energy (n.d.-a). http://energy.gov/eere/amo/static-sankey-diagram-nonprocess-energy-us-manufacturing-sector. Accessed 8/8/2016)

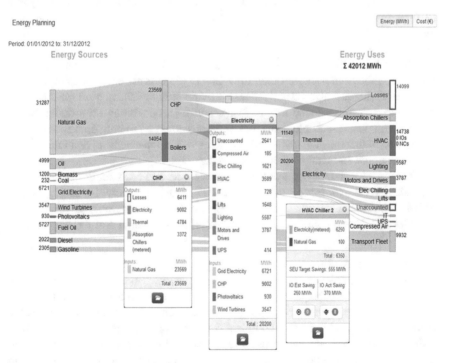

Fig. 6.4 Energy Flow Assessor. (Enerit (2017). http://enerit.com/our-software/enerit-energy-flow-assessor/. Reprinted with Permission)

for a manufacturing facility in Fig. 6.4. This type of diagram demonstrates how sensors and software can highlight relationships that were just not visible before; it also captures the results of a waste heat recapture project which reduced the waste to 27%.

Water flows inside a facility tend to follow the same types of patterns, with waste that can be recaptured, financial systems which account for water and sewer use only in retrospect (indeed, since some water utilities bill quarterly, the accounting can be even further in the past), and the purchase of water to provide secondary benefits such as cleaned vegetables or healthy employees. Water can also be purchased for its primary benefits, however, to make beer or dishwashing liquid, for example. Water flows are also closely connected to energy flows and can be measured and pictured in the same way.

6.5   A Picture of the Facility

One of the foundations of traditional energy management is the energy audit, an unfortunate term in the business context. The audit is designed to collect information on energy use in the facility, identifying areas of waste or rejected energy at a particular point in time. The audit provides information that is necessary for both understanding and addressing excessive or inappropriate energy use. There are limitations to this standard methodology, because energy use is rarely constant or predictable; where there are significant changes, the traditional audit may no longer be accurate by the time decisions regarding energy projects are being considered.

Water audits are also important and provide similar information on the water flows into a facility, inside that building, and then out through the drains to the sewer. There are several elements and questions common to both energy and water audits, and in fact they are sometimes completed simultaneously so that interactions between energy and water can be better understood.

There is no standard audit process or procedure, and each auditor may have an individual style and approach, but ASHRAE (a global society advancing human well-being through sustainable technology for the built environment) recognizes three levels of energy audits by the level of effort they entail and the outputs they generate.

Phase I Energy Audit

Also called a simple energy audit, an energy study, a preliminary audit, or an assessment, these are frequently provided at no cost by manufacturers, utility energy efficiency program personnel, energy service companies (ESCOs), and some consultants. They are generally completed in a few hours, and the client may receive a simple document highlighting basic recommendations. Generally, a Phase I audit would provide a list of no-cost/low-cost recommendations and some ideas for future planning or would focus on implementing the solution provided by the manufacturer or representative conducting the audit. Input for a Phase I audit would consist of utility data for 13 months and basic building and organizational information, possibly followed by a brief on-site inspection. If sufficient information is available, the building being audited may also be benchmarked against other similar buildings.

Phase II Energy Audit

The Phase II audit includes the information mentioned for Phase I above but goes significantly further. On-site analysis is more complete, including collecting equipment and run-time data, building occupancy information, detailed building characteristics, maintenance information, detailed utility usage and demand information, etc. This type of audit results in an in-depth look at utility components and use over time, as well as recommendations for energy efficiency measures (EEMs) that align with the client's strategies and financial realities.

Phase III Energy Audit

The Phase III audit is often called an investment grade audit because the information collected and recommendations supplied are technically and financially robust enough that contractors could conceivably use them to bid a project and banks or other financing institutions could use them to make a financing decision.

Utility Bill Audit

Another type of "audit" can be performed by a consultant or product/service provider remotely. The utility bill audit analyzes only the bills, looking for billing errors, metering errors, rate class misallocations, etc. Where these are discovered, savings can be achieved without effort from management. In deregulated environments, such audits may also be conducted by energy brokers or related parties who compete to provide lower rates for the commodity supplied by the utility (water, electricity, natural gas, etc.). In industry terms, this sort of audit focuses on the supply, not the demand.

Energy Audit Types: Which Is Better?

The level of effort and cost varies widely from one type of audit to the next, and the management's first task is to determine what they want from an audit.

In Case A, pumping equipment is critical in a chemical manufacturing facility. One pump has failed, and since the others are of similar age, an assessment from several equipment suppliers might be sufficient. Each would summarize the costs and benefits of replacing all pumps simultaneously, and some might include a much more efficient design including variable speed drives. The costs and benefits should be calculated over several years.

In Case B, a large hotel is considering a significant building upgrade with new air handling equipment, lighting, guestroom, bath fixtures, and restaurant equipment (among other projects). It would be appropriate for management to consider an investment grade audit to entice offers from energy efficiency lending institutions which design their loans to increase clients' monthly cash flow, covering the principal/interest or lease payments with monthly savings.

(continued)

> *In Case C, that same hotel might have one long hallway with rooms open-ing off on both sides. In winter, guests often complain that it takes a very long time for hot water to reach the rooms at the end of the corridor and they let the shower or sink faucets run at full force until the water becomes hot. If the hotel is located in an area of ongoing drought and the hot water heater is new and has sufficient capacity, the management would probably concentrate on addressing the problem of water use rather than that of equipment replace-ment. One way to do that would be to install a recirculating pump near the far end of the hallway to keep warm water flowing in the pipes. Most such pumps require very little power, so with this solution both water and power usage would be reduced.*

All energy audits should (but many do not) include a plan to measure and verify the savings over several years, a strategy which then enables management to prove savings, improve projects to increase savings, concentrate in other areas, or other-wise continuously drive utility costs lower.

For more detailed information, a good resource is "A Guide to Energy Audits" by the Pacific Northwest National Laboratory's Building Technologies Program in 2011[1]. There are also many books on the subject.

6.6   Real-Time Data: Monitors and the Internet of Things

Energy audits, as noted above, capture a snapshot of a facility at one point in time, with approximate calibrations for differences in daily, weekly, or seasonal usage. It is said that formulating an energy strategy using months-old data and an energy audit is similar to driving a vehicle along a winding country road in heavy traffic and changeable weather, by looking in the rear-view mirror. The analogy is apt.

Large enterprises, particularly manufacturing facilities, have enjoyed the advan-tages of control systems displaying relevant real-time data from throughout the facility including room temperatures, vent valve status, and room occupancy. These energy management systems or automated control systems are generally automated and programmed to provide comfort while being efficient in their energy use. As a result of developments having to do with the Internet of Things, even small enter-prises can enjoy the advantages of cost-effective and easily deployed automated control or monitoring systems. Both monitoring and control systems can take the place of an energy or water audit, providing real-time data and estimating costs or usage into the future, sending alerts when equipment appears likely to fail, performing comparative analyses, and so on. They enable the SEE Framework by

---

[1] US Department of Energy Energy Efficiency and Renewable Energy. (2011). http://www.pnnl.gov/main/publications/external/technical_reports/pnnl-20956.pdf.

providing accurate information that was not available previously[2] and will be described in more detail in Chap. 9.

6.7 Energy and Water Management

Even where utilities represent a small proportion of total expenses and are not part of the cost of sales, cost reductions can have significant impact on an organization's financial results. For example, a $1 reduction in water utility cost can have the same impact as a $10 sales increase, and since $4 is wasted for every $10 in utilities purchased, the savings from reducing that waste can be made available for other important purposes. Energy waste is a luxury we can no longer afford, and with new technologies and new understandings about how people in organizations think, we can recapture those wasted funds. Energy and water management drive towards increasing the energy effectiveness of organizations with information and understanding and based on a plan.

For several decades, energy efficiency products and energy conservation methods have been adopted by enterprises around the world. As those same enterprises become aware of or cope with the water risk, the importance of water efficiency becomes clearer.

From a cost perspective, water efficiency can save money on both potable water purchase costs and sewer disposal fees. The example in Table 6.1 is adapted from

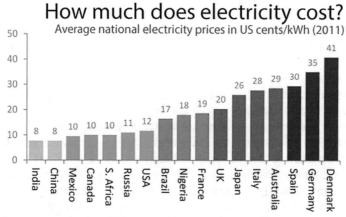

Fig. 6.5 Average national electricity prices. (Wilson (2013). http://www.theenergycollective.com/lindsay-wilson/279126/average-electricity-prices-around-world-kwh. Accessed 7/17/2017)

---

[2] Laitner (2012).

Table 6.1 Water efficiency example

Water project – typical 100,000 sq. ft. office building, 650 occupants	
Average water use/occupant/day	20 Gallons
Total annual building water use	1,250,000 Gallons / 4345 HCF (= 100 cubic feet)
Total water cost per HCF/total sewer cost per HCF/total water-related costs/HCF	$1.44 / $1.93 Total $3.37
Annual total water-related costs	$14,643
Annual total water-related costs after 30% reduction in usage	$10,249

Public Technology Inc./US Green Building Council (1996)

the 1996 US Green Building Council's Sustainable Technology booklet and high-lights the important point that water use reductions impact the charges for potable water and for sewer. In this case, the total cost reduction is $3.37 per HCF, much greater than the $1.44 per HCF price for potable water only (Table 6.1).

6.8 Utility Energy Efficiency Programs

Utility energy efficiency programs are another important and varied element of the energy efficiency landscape. Some utility programs rely on outputs from audits or energy management systems, and others provide support for such studies and analyses. In the USA they date from the 1970s, a period of high energy prices, and have become an important way for utilities to meet energy demands cost-effectively. Illogical as it may sound, utilities provide funding for and support energy efficiency programs not only because regulations require them to do so but also because energy efficiency is the least expensive way for them to meet their obligation to provide energy wherever and whenever it is required. Funding for these incentives comes from ratepayers, and it is then deployed to partially fund energy efficiency projects for those customers who request such funding.

Many programs are managed by third-party contractors, and they are an important resource for enterprises interested in controlling energy use and cost. Incentives for solar installations, electric vehicles, and other similar programs are also available in some service territories. In addition to increasing utility generation capacity, these programs reduce greenhouse gas emissions, save customers money by helping reduce their usage, and generate jobs in the efficiency/upgrade/equipment/alternative energy system installation industries.[3] Tax incentives are also frequently used to

---

[3]ACEEE, American Council for an Energy Efficient Economy. (n.d). http://aceee.org/portal/programs.

encourage energy efficiency or alternative energy investments, because they help grow the economy locally and increase the viability of local companies. In the USA, an updated database for each state can be found at www.Dsireusa.org.

Utility energy efficiency programs are designed differently in each state and by each utility; they are also regulated by the state's public utility regulation commissions. In each utility's territory, customers use energy differently in aggregate, and the generation and transmission/distribution systems vary, with attendant differences in costs. The International Energy Agency collects monthly energy statistics from countries around the world, available at http://www.iea.org/statistics/. Figure 6.5 shows a subset of that data for 2012 with huge variation from $0.41 / kWh in Denmark to $0.08 in India. The rate for the USA is $0.12, and if an adjustment were made for the average family income in the USA compared to that for India or China, prices in the USA would appear to be comparatively low.

6.9 Financial Flows

Energy flows represent one set of systems inside an organization. Another set of systems relate to financial management, because the finance department pays the utility bills. In most enterprises, the finance function is completely disconnected from the functions which manage and replace equipment, maintain internal systems, and ensure occupant comfort. Energy management cannot succeed without the finance and operations departments working together; on the other hand with a connection between the finance department and those in the organization who understand, monitor, or are primarily engaged in the activities which affect energy use, significant changes can be made quickly.

Energy Flows and Financial Flows Example

Consider a small shopping center. The owner occupies one of the retail shops and leases out three others to different tenants. All tenants share a small kitchen, meeting/training room, restrooms, and a parking area which is lit at night and on weekends. Each office has individual heating and cooling equipment, but all other utilities are shared; there is one electrical meter and one natural gas meter at the building; water is provided by the city, also through one meter.

The utility flow then looks something like this, with individual decisions or default decisions to use excessive energy having no consequences (Fig. 6.6).

And the financial flow runs something like this:

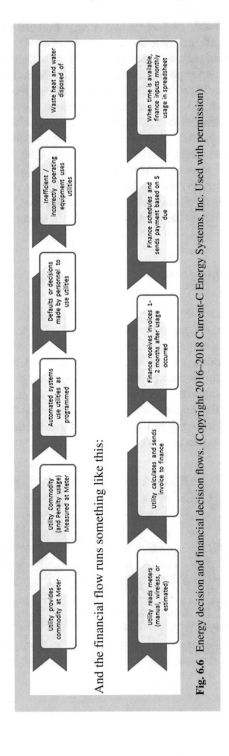

Fig. 6.6 Energy decision and financial decision flows. (Copyright 2016–2018 Current-C Energy Systems, Inc. Used with permission)

The finance department, which could engineer consequences for excessive energy use, is more concerned with paying the bills on time to avoid penalties and inputting usage and costs into a spreadsheet (which nobody looks at) whenever time permits. Since there are four different entities in the building, the situation is slightly more complicated, but even if there were only one company, the challenge would be the same: there is no connection between those who use the utilities and those who pay the bills.

Another view of this disconnect can be found in a paper by Catherine Cooremans on investment decision-making, where she sees many challenges and provides recommendations for influencing decisions. Dr. Cooremans highlights the differences between the external context and the internal one, as well as the characteristics of the actors and the investment, connecting those views through strategy. This approach can be a powerful way to understand decision-making processes in organizations, and the more tactical process flows above support that analysis (Fig. 6.7).

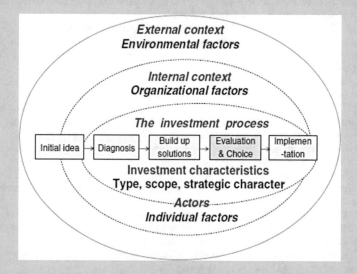

Fig. 6.7 Decision influencers. (Cooremans 2012)

A more prescriptive approach will be presented in Section III, but it is important to remember for now that in order to reduce utility costs, standard business procedures which appropriately assign responsibility and accountability are critical.

References

ACEEE. (n.d.). The state energy efficiency scorercard. Retrieved from aceee.org: http://aceee.org/state-policy/scorecard.

Cooremans, C. (2012). *Investment in energy efficiency: Do the characteristics of investments matter?* (Vol. 2012, pp. 497–518). Springer Science_Business Media B.

Diana, F. (2011, March 9). The evolving role of business analytics. Retrieved from Frank Diana's Blog - Our Emerging Future: https://frankdiana.net/2011/03/19/the-evovling-role-of-business-analytics/.

Economist intelligence Unit. (2012). *Energy efficiency and energy savings; a view from the building sector.* London: The Economist Intelligence Unit LTD.

Efficiency Valuation Organization. (2009). http://mnv.lbl.gov/keyMnVDocs/ipmvp. Retrieved from evo-world.org: evo-world.org/en/.

Eggink, J. (2007). *Managing energy costs - a behavioral and non-technical approach.* Lilburn: Fairmont Press.

Energy Star. (n.d.). Use Portfolio Manager / Understand Metrics / What Energy. Retrieved from www.energystar.gov/buildings/facility owners and managers: https://www.energystar.gov/buildings/facility-owners-and-managers/existing-buildings/use-portfolio-manager/understand-metrics/what-energy.

Enerit. (2017). our-software/enerit-energy-flow-assessor. Retrieved from enerit.com: http://enerit.com/our-software/enerit-energy-flow-assessor/.

Hansen, S. (2002). *Manual for intelligent energy services.* Lilburn: Fairmont Press.

Hansen, S. (n.d.). Making the business case for energy efficiency. Retrieved from docplayer.net: http://docplayer.net/6344792-Making-the-business-case-for-energy-efficiency-shirley-j-hansen-ph-d.html.

Laitner, J. A. (2012). *The long-term energy efficiency potential: What the evidence suggests, report number E121.* Washington, DC: American Council for an Energy-Efficient Economy.

Mourik, R., et al. (2015). What job is Energy Efficiency hired to do? A look at the propositions and business models selling value instead of energy or evviciency. Retrieved from IEADSM Leonardo Energy: https://www.youtube.com/watch?v=GGLYp_fHrMs.

New Buildings Institute. (2017, July 15). Deep energy retrofits. Retrieved from New Buildings Institute: http://newbuildings.org/hubs/deep-energy-retrofits.

Office of Energy Efficiency and Renewable Energy. (n.d.-a). Static sankey diagram non-process energy . Retrieved from energy.gov/eere/amo: https://energy.gov/eere/amo/static-sankey-diagram-nonprocess-energy-us-manufacturing-sector.

Office of Energy Efficiency and Renewable Energy. (n.d.-b). Static sankey diagram process energy. Retrieved from energy.gov/eere/amo: https://energy.gov/eere/amo/static-sankey-diagram-process-energy-us-manufacturing-sector.

Public Technology Inc./US Green Building Council. (1996). *Sustainable building technical manual.* Washington DC: Public Technology, Inc.

Shields, C. (2010). *Renewable energy facts and fantasies.* New York: Clean Energy Press.

US Department of Energy Energy Efficiency and Renewable Energy. (2011). A Guide to Energy Audits PNNL-20956. Pacific Northwest National Laboratory. Retrieved from http://www.pnnl.gov/main/publications/external/technical_reports/pnnl-20956.pdf.

US Department of Energy, Halverson, et al. (2014). *ANSI/ASHRAE/IES standard 90.1–2013 determination of energy savings: Qualitative analysis PNNL-23481.* Richland: Pacific Northwest National Laboratory.

US Environmental Protection Agency. (2017, January 24). Energy and the Environment / Electricity - Customers. Retrieved from www.epa.gov: https://www.epa.gov/energy/electricity-customers.

Wilson, L. (2013, September 25). The Average Price of Electricity, Country by Country. Retrieved from theenergycollectie.com: http://www.theenergycollective.com/lindsay-wilson/279126/average-electricity-prices-around-world-kwh.

Woodroof, E. A. (2009). *Green facilities handbook- simple and profitable strategies for managers.* Lilburn: The Fairmont Press.

Chapter 7
Energy Inside Organizations: The Four Fields

7.1 Energy Connects the Enterprise

From the point of view of a chief financial officer, financial systems are seen as the connective tissue of any organization, the link that determines what the organization looks like and how it behaves. From the point of view of an energy manager, utilities appear to be even more basic, because without them there is no enterprise. They are the "bones" without which the rest of the organization would lose form and character and the skeleton around which the organization is built. They flow everywhere.

Utilities are purchased only when – and precisely when – they are needed. They are not stored, and there is no inventory to account for. Decisions made by everyone in the organization have an impact on that instantaneous purchase, from the manager who leaves all the building lights on when working at night to the maintenance crew who keep the hot water running in the laundry area while scrubbing the floor. Utility use and costs are examined in retrospect, using bills which contain little useful information and represent the outcomes of multiple actions that happened weeks ago and will likely not be remembered by the time the bills arrive.

As previously discussed, these traditional structures and approaches mean that most organizations waste approximately 20–30% of the energy or water they purchase. It's as if the purchasing department kept only 80% of each carton of office paper purchased and tossed the rest out the window. The utility paradigm has to change, so that wasted utilities generate benefits, and the funds which were previously allocated to purchasing a wasted resource are used to accomplish other business purposes. Managers must think about utilities and organizations differently and then develop and implement plans which are practical, measurable, and relevant to the organization.

© Springer International Publishing AG, part of Springer Nature 2018
S. McCardell, *Energy Effectiveness*, https://doi.org/10.1007/978-3-319-90255-5_7

7.2   Evaluating Energy and Water Use in Organizations

Enterprises come in many sizes; at one end of the scale, they include large industrial concerns engaged in energy-intensive processes such as metal stamping, natural gas refining, plastic molding, food manufacturing, and others. At the other end are small offices with a few desks, some office equipment, and a little bit of lighting. And in between those extremes are millions of other organizations of all sizes in thousands of types of industries. It would not be possible to discuss energy use for every type or size of organization, but general rules and categories can broadly assess energy and water in any particular enterprise. This type of approach works for all types of energy including electricity, natural gas, alternative energy systems, and water, although examples are drawn from a subset of them.

   One important first step is to consider the time frame. Energy *projects* tend to be one-time investments in technologies or equipment, but they are intended to last for a much longer period. Energy *programs* tend to require less investment but are planned to continue for multiple periods. Both must be evaluated against the current situation and the projected business as usual scenario. The first part of that process consists of understanding energy and water inside the organization.

Examples: What Is the Time Frame? Short-Term Focus can Be Counterproductive

Furnace replacement:

   A nonprofit organization must replace an aging furnace. Since they have little time to devote to a detailed analysis and are chronically short of funds, they choose an inexpensive, inefficient replacement from a local handyman contractor rather than a highly efficient furnace from their air conditioning contractor. Unfortunately, although maintenance costs for the new furnace do go down, energy costs are 20% higher than they could have been, and staff complaints about comfort continue, with long-term negative personnel and financial results. The short-term focus on cash outlay to the detriment of ongoing operations probably led to the wrong – albeit justifiable – decision.

Property development:

   A property developer rushes to complete a contracted building addition, using standard construction techniques. Over the building's 50-year life, projections are that the building will cost at least 100% more to operate than if sustainable construction techniques had been used. For the developer, this seemed at the time to be an appropriate trade-off. Ten years later, however, that same developer purchased the building to operate it as an investment property – realizing at that point that sustainable construction and materials would have been a better option.

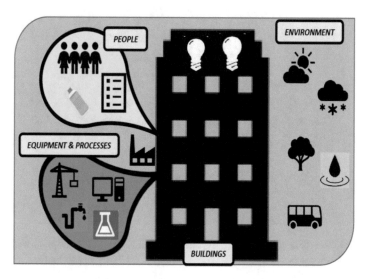

Fig. 7.1 Energy in organizations – the Four Fields. (Copyright 2016–2018 Current-C Energy Systems, Inc. Used with permission)

7.3   The Four Fields or Internal Use Sectors

In every organization, energy and water are used extensively and are more easily examined under categories, here described as the Four Fields. Since every organization is a system composed of systems, each Field is connected to the others, and all are interdependent (Fig. 7.1).

- The most obvious Field is the *building*. This includes the walls, windows, and other elements of the building envelope which separate the operation and the people inside from the exterior environment, providing the conditions necessary to conduct business. Energy is needed to heat and cool the building and to provide light and air circulation. Since buildings often last decades or centuries, they are likely to be reconfigured for multiple uses over that period, and therefore energy use patterns will change many times over the years.
- Inside the building, *equipment and processes* use energy and water in different ways.

  - Equipment consists of pumps, motors, blast furnaces, and other large identifiable single loads.
  - Processes include computers, process equipment (including compressed air systems in manufacturing operations), conveyor belts, fans, steam cleaning baths, showers, HVAC systems, refrigeration, and other systems which use energy, natural, water, or all of them.
  - Procedures are a type of process that relates to the business of the organization, not to the technical equipment; procedures are the less visible but equally important business processes and procedures including inventory control

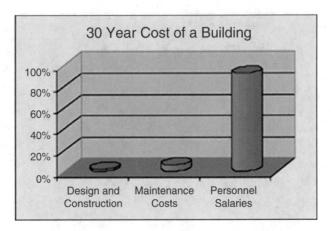

Fig. 7.2 30 Year cost of a building. (Fuller (1994, Updated 2016). https://www.wbdg.org/resources/life-cycle-cost-analysis-lcca. Accessed 3/31/2017. Used with permission)

systems, purchase order specifications, new hire training approaches, cleaning procedures, waste management instructions, and many other "soft" processes which may also connect to the other Fields.

- The *environment* outside the building also has an impact on energy use:
  - Irrigation systems keep plants alive and/or grass green
  - Heat lamps melt snow on walkways to increase staff safety
  - The sun streaming in office windows may warm – or overheat – the space
  - Transportation options might include bicycle-friendly pathways or public transportation
  - Some environments have the potential for on-site renewable energy generation.

- *People* are the last and most important Field. At one level every decision every person in an enterprise makes has impact on energy use; and on another level, only people can analyze utility use and decide how to manage them. The most critical determinants of long-term success in managing energy use are often the amount of interest the management devotes to the issue and the actions that people take to turn off computers, avoid printing unnecessary documents, and take the stairs instead of the elevator.

The Four Fields are connected in ways that are sometime surprising, as shown in the figure from the Whole Building Design Guide in Fig. 7.2. Over a 30-year period, operations and maintenance costs (equipment and processes, environment) can approximate 6%, while personnel costs including salaries and benefits (people) are over 90% of the cost of a building (building).

Enterprises vary, and therefore the balance of attention devoted to each Field will differ. Condominium offices in the center of a city will have a comparatively small focus on environment. Service industries are likely to concentrate on the buildings,

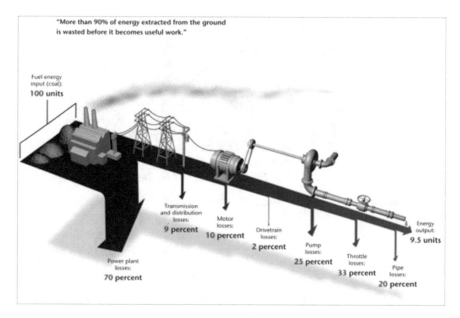

Fig. 7.3 Average energy losses in industry. ((World Business Council for Sustainable Development) 3D Sankey diagram on energy efficiency, taken from an article on "Making Tomorrow's Buildings More Energy Efficient" on the WBCSD website (http://www.wbcsd.org))

people, and equipment/process fields but most particularly on the less visible business processes and procedures. Organizations located in historic buildings may find that the building field impacts everything related to energy and water use, and so on. In other words, this conceptual model is flexible instead of fixed and should be used where it is most helpful.

7.4   Patterns of Energy Use in Different Industries

There are patterns of energy and water use shared among industries or types of organizations, as well. Fig. 7.3, the Sankey diagram of energy losses in manufacturing, shows that almost 80% of the fuel energy input is lost before the electricity is delivered to the facility. Additional losses inside the facility can be brought under management control to increase the productive use or energy output from 9.5 units. That represents a significant waste recapture opportunity.

Figure 7.4 shows power flows in a data center, a type of facility that is becoming more and more common. They use energy intensively for IT equipment, chillers, and power protection/distribution equipment, producing a huge amount of heat inside the data center and exhausting that heat as well as what is produced by the chiller.

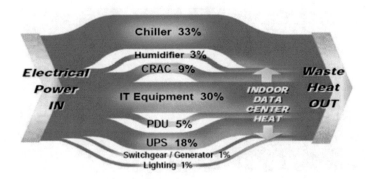

Fig. 7.4 Power flow in a data center. (APC (n.d.). http://www.sankey-diagrams.com/tag/efficiency/page/2/. Accessed 3/31/2017)

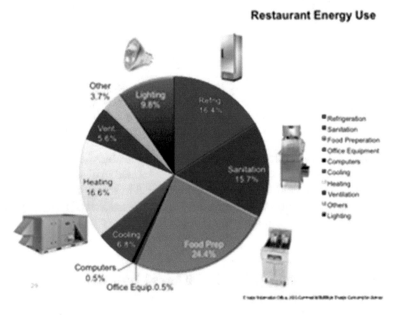

Fig. 7.5 Restaurant energy use. (Sustainable Foodservice.com (2016). http://www.sustainable-foodservice.com/cat/energy-efficiency.htm. Accessed 3/31/2017)

Figure 7.5 shows that refrigeration is key in restaurants, as are sanitation and food preparation. There are few very large energy consumers but many mid-sized ones; the total energy use intensity of restaurants is high, and the processes are inherently wasteful if they are not monitored and controlled.

Energy use intensity in retail operations, in contrast, is generally low, with most related to the building field in the room and water heating category. Lighting is both a significant cost and a critical item, because in retail establishments, the appearance and color of the merchandise may significantly affect revenue.

Fig. 7.6 Retail energy use. (Carbon Trust (n.d.). https://www.carbontrust. com/media/39228/ ctv001_retail.pdf)

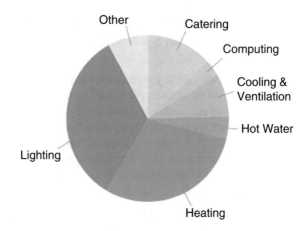

Figure 7.6 is drawn from a Carbon Trust publication entitled "RETAIL - Energy management - the new profit center for retail Businesses", a title which fits the theme of this book. A link can be found in Chap. 21, and the Carbon Trust produces informative booklets for other sectors as well.

7.5 Operations and Maintenance

In almost every facility, operations and maintenance practices and issues impact energy and water use as well as all of the Four Fields: buildings, equipment and processes, environment, and people. Good operations and maintenance practices can improve energy effectiveness, and energy effective processes and procedures can reduce operations and maintenance costs. Each enterprise will find this to be true in different ways, but some examples are:

- A practice of group relamping for the high-bay lights in a gymnasium, replacing all the lamps when a pattern of failure begins, would save energy, time required for relamping, and equipment rental costs for lifts.
- Steam cleaning equipment used in preference to mops and buckets reduces water use, cleaning time, and wear and tear on furniture from spilled water.
- Equipment maintained regularly, on a schedule, is less prone to failure and should last longer, operating more efficiently throughout its life; savings would come from reducing equipment purchase cost and energy waste.
- Buildings with insufficient insulation, gaps around windows and vents, or ill-fitting doors can be drafty and uncomfortable as well as extremely energy inefficient. Regular maintenance and upgrades keep staff happier and healthier and reduce energy costs. They may also avoid the necessity for significant investments for years.

7.6   Energy Efficiency and Energy Conservation Services, Products, and Systems: A Sampling

As noted in Chap. 4, energy use can be lowered by turning things off when they are not needed (energy conservation) and by using equipment and processes that are inherently more efficient (energy efficiency). The number of products, services, and procedures that improve an organization's energy effectiveness is steadily increasing, and every part of an institution can benefit. Given the rapid proliferation of new products and services, it does not make sense to focus on specifics; here instead are a few categories of widely applicable products or services and a few unusual or intriguing approaches that might be considered for application.

- New lighting technologies and techniques, particularly those using LED (light-emitting diode) technology are rapidly being developed. LEDs are now used everywhere, from keychain flashlights to automobile headlights and high-bay lighting in aircraft hangars. The technology is constantly being improved, increasing the already-long life and reliability of the lights, improving their color-rendering capacity, adding to their health benefits compared to other lighting technologies, and reducing their costs. Lighting systems are becoming more flexible and controllable and should always be considered for an energy efficiency upgrade, even when a lighting retrofit was recently completed.
- There are millions of motors in use around the world, and the standard motor is based on a technology which has not changed for decades. Given the high proportion of energy use and cost attributable to motors, much effort has gone into designing ways to make them more efficient. The state of the art is a moving target. For example, variable speed drives (which slow the motor to match the speed required, thereby dispersing less wasted energy as heat), variable frequency drives (which adjust the drive frequency to achieve the same end), and high-efficiency motors designed from the beginning with tighter tolerances and better materials can achieve significant cost and demand reductions. For specifics, please see the US Department of Energy's Motormaster tool at https://www.energy.gov/eere/amo/downloads/motormaster-tool .
- More efficient products can obviously reduce energy cost, and some energy conservation approaches can be equally effective. Setting up a system by which factory employees have personal incentives to turn off compressed air when they go to lunch can reduce usage by 20%, especially if leaks in the system are also identified. These human factor programs have been shown to reduce energy use, and at the same time, they increase employee engagement, often with almost no investment.
- Where waste heat can be recaptured, it may be used to reduce the need for natural gas and water for heating, dropping the need for additional energy. Consider a commercial laundry, for example; the electricity used to run the dryers also produces waste heat and humidity. Laundries commonly use natural gas to heat the wash water, so waste heat from the dryers can preheat the washing machine water, achieving significant savings.

- Plug loads such as computers, printers, vending machines, coffee machines, televisions, and other pieces of equipment that plug into wall outlets appear to be small power users individually, but in aggregate the power draw can be significant. Energy can be conserved by training staff to turn these pieces of equipment off at night or by installing switches that do so automatically.
- Functionally intentional landscaping is an often neglected but very useful way to reduce the impact of wind, sun, or other environmental conditions on a building and the people within. For example, shade trees cool in summer, evergreens cut the wind in winter, and permeable light-colored paving surfaces reduce storm runoff and heat radiating into a building.

7.7   Alternative Energy Services, Products, and Systems: A Sampling

Alternative energy systems should be chosen carefully to match the location and geography as well as the organization's needs and circumstances. Systems should be located near where they will be used to reduce transmission losses, which implies that any particular location will only have a few good options. Wind energy works better off the coast of Denmark or on the western plains of the United States than it would in the bayous of Louisiana or the Amazon Delta, for example; high temperature geothermal energy is feasible only in places like Iceland where there is volcanic activity; solar photovoltaic panels are admirably suited to Chile's Atacama Region, but not to the northern shores of Lake Superior in Canada. Zoning requirements generally do not allow wind turbines in heavily populated areas, and solar thermal panels provide heat but not power.

Alternative energy projects should be carefully considered, designed, installed, and maintained. Design and installation should be done by competent professionals, but the business reasons for the alternative energy system choice should first be clear; the technical decision should be subsidiary to the organizational ones.

Where possible, it can be interesting to design alternative energy to accomplish multiple goals at the same time. Passive solar window shading might be designed to form part of the building's architectural statement; solar panels on car parks could serve as shade structures as well; natural water treatment systems to recover gray water could be incorporated into the landscape design; and so on.

7.8   Water Products and Services: A Sampling

Water tends to be used in fewer ways and places than energy, which means that there are fewer products and services to address water use reduction applications. With increasing water scarcity, however, that situation may change. Water-related charges consist of several components: (1) purchase of potable water; (2) sewer charges for wastewater that flows into the drain; (3) stormwater charges, assessed based on the impermeable site surfaces, building square feet, or other metrics; and (4) fire hose charges. All except fire protection charges can be addressed to some extent.

Potable water:

- Irrigation systems, much like electrical systems, have been very difficult to manage because water bills arrive monthly or quarterly, showing past purchases during that period. Systems based on the Internet of Things with moisture sensors, flow monitors, and alerts can now provide the means to manage and control water use in real time.
- The most effective approach to reducing outdoor water use is to install waterwise plantings adapted to the local climate.
- Inside the building, all faucets, showerheads, and toilets should be low flow models. All process equipment should be similarly low flow and well maintained.

Sewer charges:

- Utilities often assume that all potable water purchased by a customer flows down the drain and into the sewer; in circumstances where that is not true (and the excess charges can be logically calculated), it is sometimes possible to have sewer fees reduced to a more appropriate level without investment.

Stormwater:

- Asphalt driveways and parking areas are contributors to high stormwater costs as well as to higher temperatures around a building due to the heat island effect, where dark pavement radiates heat into the surrounding environment. Permeable, light-colored designs are more effective.

References

APC. (n.d.). *tag/efficiency/page2*. Retrieved from www.sankey-diagrams.com: http://www.sankey-diagrams.com/tag/efficiency/page/2/.

Bertoldi, P. E. (2016). *Energy efficiency volume 09, issue 03*. Dordrecht: Springer Publishing.

Carbon Trust. (n.d.). *ctv001_retail.pdf Sector Overview*. Retrieved from www.carbontrust.com: https://www.carbontrust.com/media/39228/ctv001_retail.pdf.

Diana, F. (2011, March 9). *The evolving role of business analytics*. Retrieved from Frank Diana's Blog - Our Emerging Future: https://frankdiana.net/2011/03/19/the-evolving-role-of-business-analytics/.

Eggink, J. (2007). *Managing energy costs - a behavioral and non-technical approach*. Lilburn: Fairmont Press.

Energy Information Administration. (n.d.). *Today in energy*. Retrieved from EIA.Gov: https://www.eia.gov/todayinenergy/detail.php?id=18071.

Energy Star. (n.d.). *Use Portfolio Manager / understand metrics / what energy*. Retrieved from www.energystar.gov/buildings/facility owners and managers: https://www.energystar.gov/buildings/facility-owners-and-managers/existing-buildings/use-portfolio-manager/understand-metrics/what-energy.

Fuller, S. (1994, Updated 2016). *Resources/life cycle cost analysis LCCA*. Retrieved from Sustainable Building Technical Manual / Joseph J Romm, Lean and Clean Management: https://www.wbdg.org/resources/life-cycle-cost-analysis-lcca.

Gerarden, T. G. (2015, January). *Addressing the energy-efficiency gap faculty research working paper series RWP15–004*. Retrieved from Harvard Kennedy School: https://research.hks.harvard.edu/publications/workingpapers/Index.aspx.

Hansen, S. (2002). *Manual for intelligent energy services*. Lilburn: Fairmont Press.

Hansen, S. (n.d.). *Making the business case for energy efficiency*. Retrieved from docplayer.net: http://docplayer.net/6344792-Making-the-business-case-for-energy-efficiency-shirley-j-hansen-ph-d.html.

Irandoust, S. (2009). Sustainable development in the context of climate change: a new approach for institutions of higher learning. *Integrated Research System for Sustainability Science*, 135–137.

Jewell, M. (2014). *Selling Energy*. San Francisco: ISBN: 978–1–941991-00-8.

Public Technology Inc./US Green Building Council. (1996). *Sustainable building technical manual*. Washington DC: Public Technology, Inc.

Sustainable Foodservice.com. (2016). *Energy-efficiency.htm*. Retrieved from www.sustainablefoodservice.com: http://www.sustainablefoodservice.com/cat/energy-efficiency.htm. Accessed 3/31/2017.

Woodroof, E. A. (2009). *Green facilities handbook- simple and profitable strategies for managers*. Lilburn: The Fairmont Press.

Chapter 8
Benefits, Barriers, and Opportunities

8.1 Why Understanding Benefits and Barriers Is Important

There is an exhaustive literature on the drivers and barriers for energy efficiency and alternative energy projects, much of it written by people in the industry who discount the complexity of organizational prioritization and decision-making, focusing instead on what they see as the obvious technical benefits of engaging and investing in energy projects. Most people in the energy efficiency industry are engineers, so the industry is inclined to first consider good technical decision criteria before making a broader evaluation of what might be a good business decision. For this reason, it is important to understand some of the benefits of energy efficiency, water, and alternative energy projects, as well as some of the barriers.

Drivers or Benefits will naturally vary by type of institution, size, intensity of other challenges, maturity, goals, culture, and in many other ways. Organizations are likely to move forward when at least three benefits resonate, when co-benefits can be presented, or when an opportunity presents itself to rethink/reexamine the barriers and find a way to work around them.

Energy projects commonly affect multiple systems simultaneously, and the impacts can be positive or negative. For example, a lighting project in an office building might reduce lighting and air conditioning costs, increase heating costs, reduce glare from computer screens, improve employee health, improve security, help the cleaning crew be more efficient, allow the maintenance team to focus on other important activities, and make the building's interior look more "modern" or attractive to clients and customers. The system-wide impacts will also occur when projects are undertaken for reasons that have nothing to do with energy. For example, replacing or improving drafty windows in an older building because of employee complaints should make those employees more comfortable, reduce their need to take sick days, improve employee morale, reduce utility costs, and cut building maintenance expenses.

© Springer International Publishing AG, part of Springer Nature 2018
S. McCardell, *Energy Effectiveness*, https://doi.org/10.1007/978-3-319-90255-5_8

8.2 Categories of Benefits

For any utility-related action or project, there are likely to be three important categories of benefits, some of which are difficult to quantify. Co-benefits might form part of the same category, or fall in different ones.

- *Financial Utility Cost Reductions.* This is the easiest to measure and report and includes reductions in demand cost or other penalties as well as reductions in usage. Often, these utility cost reductions are the only "benefits" considered in energy effectiveness projects or programs.
- *Financial Nonutility Cost Reductions.* These cost displacements are more difficult to associate directly with energy effectiveness projects but include, for example, the reduction in maintenance costs from replacing an older and less efficient chiller with a new, highly reliable, high-efficiency unit. Installation of a new solar system to provide power during high-cost periods and during blackouts would reduce daily costs and avoid the necessity of closing the facility – an avoided cost which is difficult to quantify but nevertheless real.
- *Nonfinancial benefits.* These types of benefits, no less real than financial ones, are inherently difficult to quantify. They might include an advantage in attracting people who find the idea of working in a newly updated and sustainably designed building to be a draw or the positive community impact of opening company hazardous waste collection facilities to community members.

8.3 Motives to Move Forward

There has been a great deal of research on what motivates managers in organizations to say "yes" to projects which improve energy effectiveness; some analyses show that the primary motivations are, in order:

1. Avoiding risk
2. Avoiding hassle
3. Gaining praise
4. Gaining power
5. Having fun
6. Making a profit[1]

8.4 Benefits and Connections

Those managerial motivations are reworded in Table 8.1 into standard business terminology, with examples of some potential actions and an indication of the benefits/co-benefits that an organization might expect, all organized by the Four Fields.

---

[1] Jewell. (2014).

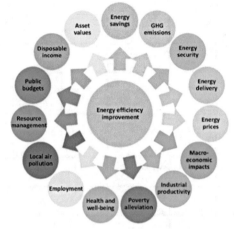

Fig. 8.1 Multiple benefits of energy efficiency improvements

Please note that "general business structure," the first column, is a general systems category which may include all others. For example, if a manufacturing company chooses to outsource their manufacturing or a product distributor chooses to bring the sales function in-house, everything related to utility use may change. The list is not intended to be exhaustive but rather to showcase potential benefits and co-benefits.

These co-benefits can be felt outside the organization as well, as shown in this slide from one of Leonardo Energy's presentations where it is noted that "Energy efficiency is a means to enhance energy security, support economic and social development, and promote environmental goals." (Fig. 8.1)[2].

Co-Benefit Example

A food processing operation is located in a drought-stricken area. The water used to wash the produce is rapidly rising in price, and the company is considering implementation of a project intended to reduce water use. One option would be to adapt their washing process to use less water and use nontoxic chemicals at the same time. The wastewater could then be recaptured and used instead of potable water to irrigate the surrounding gardens. Marketing and public relations efforts – both internal and external – could highlight the positive effects of the initiative. Staff might be inspired to submit additional ideas for waste reuse. Such a program would bring benefits across all categories, in all Four Fields.

Table 8.1 Benefits and connections for selected energy effectiveness strategies

Activity/project	General business structure	Risk reduction	Competitive-ness/no "hassle"	Support values "Praise"	Sustain-ability "Power"	People/staff/"Fun"	Cost reduction "Profit"
Building							
Building design/redesign		X				X	X
Eliminate waste and cost in bldg.		X			X		X
Offset reliability risk/downtime for utility services		X	X			X	X
Energy Star-labeled office bldgs.	X	X	X	X		X	X
Maintenance planning		X	X				X
Improve space utilization			X			X	X
Lighting quality					X	X	X
Safety		X		X		X	
Improved technology					X		X
Business processes/process and manufacturing processes							
Amend fundamental business model	X		X	X	X		
Adjust production process and design		X	X		X		X
Improve business flow and process	X	X	X		X	X	X
Improve business policies, procedures	X	X	X	X	X	X	X
Improve marketing			X		X		
Reduce waste and cost in processes		X	X		X		X
Improve Mfg reliability and quality		X	X		X		X
Process changes to reduce reg. compliance risk		X	X				X
Reduce procurement risks		X	X				X

(continued)

Table 8.1 (continued)

Activity/project	General business structure	Risk reduction	Competitive-ness/no "hassle"	Support values "Praise"	Sustain-ability "Power"	People/ staff/"Fun"	Cost reduction "Profit"
Reduce degradation of production assets, HVAC, and other systems			X		X		X
Redirect energy $$ to other purposes			X	X	X		X
Reduce technology risks		X	X				X
Further organizational sustainability				X	X	X	
Reduce "fugitive" energy (which destroys equipment, etc.)		X	X				X
People							
Attract and retain staff		X		X		X	X
Increase occupant comfort				X		X	X
Increase employee health				X	X	X	
Attract and retain tenants						X	X
Improved worker productivity				X	X	X	X
Reduce absenteeism or presentism			X			X	X
Environment							
Reduce GHG emissions				X	X		
Join actions to vision of improving community, etc.		X		X	X	X	X
Improve public perception		X		X			X
Increase s/h value	X		X	X	X		X
Increase value of enterprise	X		X	X	X		X

There are many compelling reasons to invest in energy effectiveness programs, balanced by barriers which are often business- or people-related.

8.5   Constraints and Barriers to Moving Forward

There are always potential technical barriers to energy projects, although they are not considered in this book; it is assumed that technical experts will advise management on technical questions.

The more relevant barriers are those which relate to the constraints under which every organization operates, including:

- Management time and "share of mind"
- Internal levels of expertise and available time
- Financial constraints such as funds available for investment, operating funds, or borrowing capacity
- Lack of understanding about how utilities are used, the impact of that use, and approaches to managing or controlling them

These are true constraints, appropriately handled by managers experienced in choosing between available alternatives in a resource-constrained world as part of their discipline. Projects intended to increase energy effectiveness should properly be included in the options considered by management for organizational growth and strengthening. The list of barriers below is selected from the literature and is included to frame the analysis.

- Building:

  - Leased rather than owned
  - Technical complexity
  - Permitting issues
  - Lack of data and information

- Equipment/Process:

  - Complicated potential financing sources
  - Difficulty determining appropriate method for analysis
  - Identifying inefficiencies

- People:

  - Lack of communication across divisions
  - Inadequate understanding of benefits
  - Split incentives with investor not receiving benefit/credit for savings
  - Inertia

- Environment:

  - Inadequate information dissemination from product and service providers

- Difficulty finding trustworthy/impartial sources
- Uneven, inconsistent, and unpredictable legislation, policies, incentives
- Unclear distribution channels for products and services

8.6 Leadership, Accountability, and Cooperation

Without both leadership and accountability, increasing energy effectiveness is unlikely, whatever the benefits of doing so might be. Leadership moves consideration of energy issues to a high priority where resources and management attention are carefully allocated: upper management and middle managers must all be involved in utility management; middle managers and line managers have much to contribute and can become roadblocks if they are not involved.

Projects cannot be accomplished without accountability, which is only possible once there is enough information for managers to understand what they are accountable for. For example:

- Energy cost control requires cross-departmental planning, cooperation, and leadership to be truly effective, and organizations do not normally work that way. A compressed air system, for example, uses 5–7 HP of electricity to generate 1 HP of compressed air.[3] Improvements to that system may have the potential to drop the utility bill from $500,000 to $400,000, but in order to accomplish that reduction, the workers on the factory floor, the maintenance and operations people, the finance department, and the production engineers (and likely others as well) must all be involved in planning, designing, implementing, and evaluating the project, or it is likely to fail.
- Budgeting in many enterprises is done by division, and energy projects rarely fit cleanly within one division. Where departments are tempted to hoard their budgets, it is difficult to see where energy projects can fit and therefore be approved.

8.7 Other Impediments to Action

Myths, common belief sets, and rules of thumb can also be barriers to action. Some myths have never been accurate, others are outdated, and still others are in place more as default beliefs than for any logical reason.

Myths:

- Energy is an uncontrollable fixed cost.

---

[3] Russell. (2010) p. 60.

- *In fact utilities are usually both variable and manageable. Energy and water use and penalties should be related to particular activities or products and tracked and compared to key performance indicators (KPI), in the same way as material costs or personnel hours are. Deviations from the norm should be analyzed and corrective action taken.*

- First cost or investment is more significant than ongoing costs

  - *For vehicle purchases, the purchase price is one consideration, along with operating costs, reliability/maintenance and repairs, and other ongoing before the vehicle's anticipated sales date. When most businesses decide to purchase energy-related equipment, they consider only the first cost even when the reason for the investment is to reduce costs.*

- We've always done it that way or don't fix what ain't broke

  - *People have a tendency to assume that the current situation is static. Outdoor lighting control timers are configured at installation and then forgotten. Water heaters, notoriously subject to corrosion and damage from hard water, are assumed to operate at the same level of efficiency until they fail. Office buildings are expected to function as they always have, even if the interior wall configuration has been changed 3 times in the 15 years. In each of these cases, utility use can increase dramatically – and it does.*

- Efficiency = environmentalism; conservation = environmentalism

  - *It is interesting, yet counterproductive, that the terms "environmentalism," "conservation," and "efficiency" have taken on political overtones, with resource conservation not being considered conservative in the utility field. Utilities are only one of many resources that organizations conserve and balance – along with time, funds, customer satisfaction, and other elements.*

- We don't have enough money

  - *For many investment decisions, postponing the capital outlay is the conservative choice, and making do saves money. A building owner might decide to upgrade the office's exterior by painting the building herself, rather than installing a new façade. A property manager might just add additional gravel to the parking lot, instead of paving it. These may be prudent business decisions and save funds for other purposes. With energy, on the other hand, spending money can sometimes be the prudent decision, even if it is necessary to borrow money to make the required investment, because the net savings improve the organization's cash flow situation. The total savings that will forever be lost if the decision is postponed are referred to as the <u>Cost of Delay.</u>*

- We just replaced that, so I'm not going to spend more money on it

  - *The <u>Sunk Cost Fallacy</u> is a temptation familiar to most business people; the term captures the unwillingness to spend more money to replace a recent project. The question of installing a solar PV system on a building whose roof* *was replaced 2 years ago, for example, should be examined on its merits given the current situation; the funds used to replace the roof have already been spent. Yet the temptation to bring the roof installation into the analysis is strong.*

Outdated beliefs:

- Funds for capital improvement projects are not available

  - *Over the last few decades, several institutions have created financing products specifically for energy efficiency projects because they see where their repayment is going to come from – it is the increased cash flow. There are multiple financial products appropriate for varied circumstances.*

- Waste is waste; it should just be disposed of

  - *Wasted energy is created in all processes (as noted in earlier chapters), but it does not then need to be disposed of. Wasted energy can sometimes be recycled in the same way that paper can, being used to provide new benefit. Good energy programs aim to reduce waste first and then if possible to recapture the waste for further use. One example of this paradigm shift to seeing waste as a resource can be found at dairy and hog farms, which use plentiful animal waste to generate power and heat.*

- Fear – Does this mean I've done a bad job?

  - *Many projects are never proposed because of fears that doing so will indicate the maintenance person or facility engineer has fallen behind in keeping the operation energy efficient. The fact is that ALL operations are inefficient to one degree or another, and the person responsible for energy projects should continue to search for improvements.*

Inadequate understanding/default thinking:

- Nobody ever saves his/her way to the top

  - *Financial and manufacturing professionals can, and in fact often do, "save their way to the top." The same should be true of those who implement energy programs, which also produce savings that go straight to the bottom line.*

- Incomplete understanding of financial benefits and life cycle cost

  - *Energy-related decisions are often made after considering the payback period (investment divided by the yearly savings) rather than the more complete life cycle cost methodology. It is just not an appropriate project analysis tool.*

- The finance department just pays the bills

  - *The most common pattern is for the finance department to pay utility bills without asking departments to account for, analyze, or explain their utility usage so usage and cost are disconnected.*

- Convenience

  – *It IS convenient to have utilities delivered where we want them, when we want them, in the quantity we need, and without further thought. On the other hand, further thought might help accomplish other organizational objectives.*

- Looking forward through rear view mirror

  – *Utility costs are ignored because there has been no data on past or projected spending that is actionable. Organizations drive forward, looking through the rear-view mirror.*

8.8   Strategic Utility Management

Investments in energy effectiveness should relate in a strategic way to the organization's basic business purpose. This connection provides an additional screen through which project proposals must pass as well as the potential for co-benefits that come from coherence between all parts of the enterprise.

- A food bank should consider investing heavily in energy efficiency if their mission is to distribute food to people who need it and to rescue food that might otherwise be wasted; reducing waste in other areas would be consonant with their mission.
- A nursing home is committed to keeping their residents comfortable and healthy. Costs for such facilities are increasing, so the most efficient and flexible space conditioning systems that are also cost-effective to install and operate over 10 or more years would be an excellent investment, as long as the systems exceed health regulations.
- A school dedicated to educating young people might teach them to be good stewards of resources by installing and explaining low-flow faucets in the sinks, making students responsible for checking to see that all lights and computers are off at the end of the day, or having a Vampire Contest to discover devices that use power even when they seem to be off.
- A supermarket relies on rapid turnover and has very thin margins, so good quality lighting is important and every penny saved in operations affects the bottom line; the supermarket should invest in the latest lighting technology to reduce costs and improve the food's attractiveness.

References

Chiaroni, D., et al. (2016). Overcoming internal barriers to industrial energy efficiency through energy audit; a ase study of a large manufacturing company in the home appliances industry. Clean Technology Environmental Policy.

Economist intelligence Unit. (2012). *Energy efficiency and energy savings; a view from the building sector.* London: The Economist Intelligence Unit LTD.

Energy Information Administration. (n.d.). Today in Energy. Retrieved from EIA.Gov: https://www.eia.gov/todayinenergy/detail.php?id=18071.

Energy Star. (n.d.). Use Portfolio Manager / Understand Metrics / What Energy. Retrieved from www.energystar.gov/buildings/facility owners and managers: https://www.energystar.gov/buildings/facility-owners-and-managers/existing-buildings/use-portfolio-manager/understand-metrics/what-energy.

Fry, A., et al. (2005/Reprint 2006). *Facts and trends - water.* London: World Business Council for Sustainable Development, wbcsd.org.

Jewell, M. (2014). Selling Energy. San Francisco: ISBN: 978-1-941991-00-8.

Russell, C. C. (2010). *Managing energy from the top down; connecting industrial energy efficiency to business performance.* Lilburn: The Fairmont Press Inc /CRC Press.

SustainAbility. (2014). 20 Business Model Innovations for Sustainability. SustainAbility.

The Shift Project. (2012). Top 20-Capacity Chart. Retrieved from http://www.tsp.org The Shift Project: http://www.tsp-data-portal.org/TOP-20-Capacity#tspQvChart. Accessed 2/17/2017.

Thorpe, D. (2013, July 1). Demand side response: Revolution in British energy policy. Retrieved from The Energy Collective.com: http://www.theenergycollective.com/david-k-thorpe/244046/demand-side-response-revolution-british-energy-policy

US Energy Information Agency. (n.d.). faqs. Retrieved from eia.gov: https://www.eia.gov/tools/faqs/faq.php?id=427&t=3

US Environmental Protection Agency. (2017, January 24). Energy and the Environment / Electricity - Customers. Retrieved from www.epa.gov: https://www.epa.gov/energy/electricity-customers.

Chapter 9
Tools and Methodologies

9.1 General Comments

People are comfortable with what is familiar. They just like to have it presented in new ways. In that vein, the three particular methodologies and tools upon which the Strategic Framework presented in this book rests are familiar to managers and financial officers (financial analysis), to engineers and developers of computer applications (real-time data and the Internet of Things), and to social scientists and economists (human behavior/institutional change methodologies). The tools and methodologies are:

1. Financial analysis: This approach is familiar in all enterprises, applied somewhat differently to energy projects. Financial tools should be used to evaluate projects in advance and to evaluate results.
2. Real-time data: Circuit miniaturization and wireless capabilities provide the opportunity to see energy flows as they happen, so that managers and staff can take action in real time.
3. People: People can turn an energy program into a success, or they can doom it to failure. At times, they can even BE the energy program. Behavior change techniques should form part of any energy project or program.

The last set of tools in this chapter do not relate directly to the framework, but they do enable projects to move forward; they are potential sources of funding for energy projects.

9.2 Financial Tools and Methods

Managers track and analyze most activities using financial metrics which knit together the relationships inside the organization. Utilities are also intimately connected to every other part of the enterprise; a similar approach should be used with

© Springer International Publishing AG, part of Springer Nature 2018
S. McCardell, *Energy Effectiveness*, https://doi.org/10.1007/978-3-319-90255-5_9

utilities. As discussed previously, however, energy investment and energy savings decisions are sometimes considered and evaluated by different criteria, leading to inappropriate decisions. Management could be seen as the ability to appropriately allocate scarce resources to ensure that the enterprise thrives; those resources to be allocated include electricity, water, and natural gas no less than people, cash, or copy paper.

9.3   How to Evaluate Energy Projects

After acquiring a conceptual understanding of energy use in their organizations, managers should consider intangible elements including its purpose and values to help inform that process, reducing costs and making a statement about the organization. Energy effectiveness should be accomplished strategically, with more detailed technical analysis to follow after the strategic questions are addressed and the initial financial assessment made.

The initial analyses of an energy project should be made using the organization's standard financial process, not the simple payback period often employed for energy projects which can be misleading (although it is simple.) For organizations comfortable with using a "hurdle rate," that is the appropriate measure. For those who prefer return on investment, that is the correct approach. Where other metrics are standard, those should be used. Only with comparable metrics and analyses can energy projects be compared to other potential uses of funds.

Whatever the calculation, the comparison should be to a base case which is normally the status quo. The base case is normally that which currently exists carried forward into the future and including projections such as increased maintenance costs as the equipment ages, disposal costs or value at end of life, utility cost projections, parts obsolescence, etc. The proposed case, or cases, should of course include the same categories of projections as well as the investment required, financing methodology and associated costs, discount rate, equipment life anticipated, etc. Generally, the assumptions and projections regarding utility price and usage, maintenance frequency and cost, parts, etc., will be different for each case examined, so this type of robust analysis can provide management with a good decision-making matrix.

Alternative Energy/Business Metric Example
Always use the organization's standard decision-making analyses. For alternative energy investments, the appropriate analysis might be one generally used for Make vs. Buy decisions. In this case, a large nonprofit is considering installing a solar system on shade structures in part of the garden near the parking lot to provide peak energy use reduction and backup power as well as a place for employees to meet for lunch.

(continued)

Using a Make vs. *Buy analysis, the company is buying its electricity at an approximate yearly cost of $1,000,000. Price increases are projected at 6% a year, and usage is projected to be flat. Forty percent of total cost is related to peak demand penalties, which occur in the afternoon (peak production time for a solar array).*

The alternative under consideration is to install a solar array costing $1,500,000, to be financed at 10% over 10 years. That solar array should cover 100% of peak demand charge and 50% of the total utility usage cost with no increase in price. Price increases would still apply on 50% of total usage, and maintenance costs for the solar array will be $50,000 per year. Depreciation is not accounted for. Over the 10 years the organization anticipates remaining in the building, which option is better? (Table 9.1)

The analysis above assumed that the installation of the solar array was a straight equipment or property loan, but in fact there are many ways that energy efficiency and alternative energy projects can be funded or financed. From this simple analysis, it appears that the solar system would reduce electricity costs over the 10-year period from $13,180,795 to $7,279,238 making it the clear winner despite the $1,500,000 installation cost.

9.4 Financial Methodology Challenges

The Color of Money
One of the challenges in most organizations is that money is not treated as the commodity it is, but looked at in different ways depending on where it is accounted for and the purposes for which it is intended; it seems to have a particular color.

Color of Money Example
Beta Manufacturing has an operating budget of $100,000 for the next budget cycle. The responsibility for control of this budget item rests with the Director of Operations. The capital investment budget for the next year is $10,000, and that budget is controlled by the Director of Engineering.

The Engineering Manager presents to the Director of Operations a plan to install heat reclamation equipment for the natural gas furnaces in order to reduce natural gas costs. The investment cost for this project is projected to be $12,000, and annual operating cost reductions are projected to be at least $7000. Company management decided to delay the project indefinitely because there were insufficient funds in the capital budget.

Table 9.1 Comparison of Make vs. Buy Options

	"Buy" Option / Current Situation		"Make" Option, Solar System	
	Annual	10 Years	Annual	10 Years
Total electricity cost, 6% increase	$ 1,000,000	$ 13,180,795	$ 300,000	$ 3,954,238
Solar investment Total			$ 1,500,000	
Yearly payments, principal & interest			$ 300,000	$ 2,825,000
Maintenance on solar system			$ 50,000	$ 500,000
Total outflow, "make" option			$ 650,000	$ 7,279,238

It is difficult to argue that the decision reached by Beta Manufacturing was the correct one for the company as a whole, though perhaps it allowed the Director of Engineering to save money for another pet project. With energy efficiency projects, the amount that is in the budget is one issue, but another is that those who approve energy efficiency investments or energy efficiency programs may not be those able to either evaluate the project's impacts or take advantage of them. In a hospital, for example, the engineering department would research and specify a new chiller, the finance department would approve the purchase and installation, and the facilities department would be saddled with the result – an effective chiller that was easy to maintain and kept energy and other operational costs to a minimum, or a less costly piece of equipment (up front) that required monthly maintenance for reliability and missed the opportunity to reduce energy costs.

Sunk Cost Fallacy

One of the tenets of financial analysis is that when comparing alternative uses for scarce resources, logical alternatives should be described and quantified. Since enterprises are frequently engaged in multiple programs and projects, and since utilities are integral to all operations, it is almost inevitable that managers will be faced with a situation that involves the Sunk Cost Fallacy and have to make the effort to focus on projections of the future without considering the residue of decisions previously made. For example, the owner of a small strip mall just repaved the parking area in response to tenant complaints. Six months later, the city announced incentives for replacing asphalt in parking lots with pervious paving stones in order to recharge the groundwater resource. In making a decision about replacing the new asphalt with pervious paving stones, the building owner should properly ignore the sunk cost of repaving and focus on the completed parking lot as the base case, with the investment option being the impervious paving project including city incentives.

Cost of Delay

Where utility projects are delayed for technical, operational, management, or other reasons, the impact on the organization may be substantial. Inertia and other priorities may cause delays which, in the case of energy projects, means that the organization may continue to bear unnecessary costs. These are quantified in a calculation known as the Cost of Delay.

Cost of Delay Example

Consider an office building which is considering replacing its old fluorescent lights 1 for 1 with LED bulbs. The maintenance and engineering staff interview the contractors, choose the bulbs, and give an analyst in the finance the task of analyzing the project. The analyst decides that, for comparison purposes, it would be inappropriate to include all the other details normally required by the finance department such as anticipated changes in maintenance costs, expected utility rate changes, cost of capital, projected inflation rate, and others; the analysis is capped at 5 years despite the fact that LED bulbs should last at least double that amount of time.

Current Situation:

- *100 fixtures, each 2 lamp T12 FL with ballast; total W per fixture = 165*
- *kW for lighting 16.5 for 11 hours/day or 4016 hours/year = 66,264 kWh*
- *Electricity cost for current fixtures at 11c/kWh = $7289.04*

Replace Lamps and Ballasts 1 for 1:

- *100 fixtures, each 2 lamps LED, total W per fixture = 36 kW*
- *kW for lighting 3.6 for 11 hours/day or 4016 hours/year = 14,457.6 kWh*
- *Electricity cost for new fixtures at 11 c/kWh = $1590.34f*
- *Cost for installation $17,000 including materials and labor*

Annual Electricity Cost Savings/Analysis:

- *Annual electrical savings per year with new fixtures: $5698.70*
- *Year 1 cash flow $(11,301.30); cash flow for years 2–5 $5698.70*
- *Internal rate of return (calculated by Excel) at 35%*

Knowing that an ROI of 35% is excellent and far greater than management would normally anticipate, the analyst also decides to calculate the cost of delay:

Cost of Delay:

- *Without accounting for the investment cost, savings lost per month would be $475 at least.*
- *Including the investment and the yearly savings as in the previous analysis over 5 years, the average cost of delay per month is $192 for 60 months.*

With this analysis in hand, management realizes that it does not pay to wait.

9.5   Organizational Behavior and Human Factors

Energy effectiveness can be accomplished in several ways. For decades, that effort focused on the buildings (green buildings, net zero buildings, natural building, and other strategies) and on the equipment in those buildings (LED upgrades, high-efficiency boilers, new insulation compounds, solar generation, and so on).

Many consultants and engineers can design and specify energy efficiency projects or systems chosen specifically for a particular building or process; energy efficiency is a technical specialty about which there is a great deal of information available. A selection of those resources can be found in Chap. 21.

In addition to the technical energy efficiency projects, there are other types of programs which can be conceived, designed, and carried out by the business managers themselves using familiar methodologies in a new way.

With monitoring systems and new ways of understanding how people make decisions, managers now have nontechnical opportunities to reduce energy use and cost as well as the ability to track those changes and adjust where necessary. For the first time in the history of wide-scale utility use, it is possible for even smaller organizations to design programs, set up feedback loops to measure results, and tune their processes to improve energy effectiveness.

Utility use and cost in a facility result from hundreds or thousands of actions and decisions people take every day by default or with intention. In the new arena of managing to improve energy effectiveness, several trends have come together simultaneously; the extent to which people's decisions and actions matter is only now becoming clear because of several innovations, applications, and discoveries:

1. Frameworks for organizational change are being researched and better understood, with multiple methodologies being developed to assist in the change process.
2. The fields of cognitive science and neuroscience are making substantive discoveries about how people think and make decisions.
3. It is becoming more of a societal norm for people to pay more attention to the importance of their individual actions in reducing energy use.
4. The "Internet of Things" has led to miniaturization and communications which affect what we can learn about energy use in real time, and over time.

Behavioral economics addresses the reasons people make decisions which appear to be irrational from the point of view of classical economics, a discipline which asserts that people are inherently rational and make totally rational decisions based on marginal benefit and cost. Through creative and fascinating experiments, those active in the field of behavioral economics have developed some principles which explain why people behave in predictable ways even when those ways are not logical. These principles also help explain why energy effectiveness is so difficult to achieve.

Principles of Behavior Economics
- People value avoidance of loss more than the promise of gain.
- People make comparisons, not decisions; comparison should be framed carefully.
- In the absence of math, decisions are made by myth.
- Emotional appeal is more important than technology.
- Motivation is more important than education or awareness.
- People are motivated more by relative values than by absolutes.
- Immediate effects are important.
- Resistance to coercion is common; to move forward, humanize the task; give positive reasons.
- Guilt does not work.

These principles help explain who we are as human beings – and how we can improve our energy effectiveness. Despite economists' decades of publications and assumptions to the contrary, behaviors are rarely rational; people (including management and staff) are more productive when they feel safe, secure, accepted, loved, free, and in control. Cultural identity and social networks, relationships, affiliations, and emotional needs are extremely important. Knowing this, marketers and utilities have used competition, social networks, and technology to increase recycling rates in certain cities, for example, or to compare energy use between neighbors.

In an organizational context, the same types of approaches can be designed and implemented; the list of suggestions for that design is summarized from successful programs:

Design Components for Successful Human Factor Programs
- Information should be emotional.
- Leverage comes from purpose, mastery, and autonomy.
- Incentives as a prize do not stimulate lasting change, which requires a tapestry of functional and emotional benefits.
- Highlight the cost of delay where relevant.
- Avoid overcomplication.
- Energy awareness is not enough, but it can be the first step.
- Conversations convert – ask questions, get feedback, do not persuade but engage.
- People are different; what motivates one may not motivate others. Ask and listen.
- People want flexibility and control.
- Feedbacks – devices, communication, platforms – make behavior or the consequences of behavior visible.
- Reinforce the good and make it fun.
- Remember that change is multifaceted and can take a long time.
- From the Five-Step Framework, there are eight principles to help programs be effective:
  - Use social networks – seeing or hearing of others behaving differently.
  - Social empowerment – feeling a part of something larger.
  - Social commitment – making definite commitments, especially publicly.

- Information and feedback – receive actionable information and feedback.
- Multiple motivations – reacting to varying interests.
- Leadership – seeing leaders' support at all levels.
- Infrastructure – change to make new behaviors easy/desirable.
- Continuous change and innovation – it is a process not an outcome.

• Participants can be engaged by using some of the following types of initiatives, among others:

- Workshops
- Treasure hunts (for energy or water waste)
- Train the trainer programs
- Competitions
- Cooperative projects

These programs are generally long-term and require continuous commitment. In the literature, the term of art is behavior change, and that does not consist of creating awareness, increasing knowledge, or changing attitudes. Those are all important and sometimes helpful, but behavior change is intended to change actions whether or not awareness and knowledge have increased, or attitudes have changed.

With that in mind, the Strategic Framework around which this book is based uses the term energy management or human energy management to distinguish it from the automated control systems sold by Honeywell, Siemens, and others which are designated as EMS, energy management systems. The principles and methodologies developed by those engaged in behavior change are core to this type of energy management, but for managers to be able to design and implement energy effectiveness programs, they need to be aware of energy and understand how it's used.

9.6 Energy and Water Monitors and Systems

As has been noted previously, the Internet of Things brought about by the miniaturization of circuit boards and intelligent systems enables the capture of information unavailable before. Cellular phones, for example, contain more computing power than the original room-sized computers, and tiny sensors such as those designed to go inside the human body can track light/dark or anything that flows such as electricity, natural gas, or water.

In the energy field, this is important because, as has been said, "What gets measured gets managed." If a metric is available (cartons of paper, tons of food waste from the cafeteria, average length of time staff remain with the organization, minutes needed to complete a specific task, etc.), then the organization can include that data in key performance indicators and other results. Indeed, even the capability of tracking usage and costs can have a strong effect: one Department of Energy study determined that utility costs were reduced 25% or more just by metering and communicating results[1]; individuals took it upon themselves to reduce energy use and waste.

---

[1] Eggink. (2007). p.73.

Monitors enable a building owner or manager to learn how energy is used in a building or a company and work with staff to create positive change. There are essentially three types of systems which are used and useful in this context:

1. *Smart meters* are used by some utilities (whether electricity or natural gas providers) to track and report usage in real time. The utilities use that information for their own purposes, but managers may be able to access the more detailed data as well, on a historical basis. Events could not be captured or addressed as they happen, but the additional detail might provide a connection between usage and activities. For example, a surge in water use in a previous week might cause management to send a crew to find and fix a leak before thousands of gallons of water are lost.
2. *Controls systems* already installed to control the air handling and building conditioning equipment may also include view/monitor/alarm settings and collect historical data in ways that are useful to management, without additional outlays. Where there are no controls in place, flexible and reasonably priced new systems are also available, some of them able to integrate equipment from multiple vendors.
3. *Monitoring systems* are appropriate for circumstances where a human interface is more appropriate or needed. With real-time feedback and information, people can take immediate action as appropriate or collect performance data over a period of time. Some of these include:

   (a) Sensitive processes which must be flexible to accommodate unpredictable conditions, and where an automated response could be counterproductive or even destructive
   (b) Cash-poor environments where it is difficult to finance equipment or other purchases and staff are involved in technical monitoring
   (c) Situations where people want to learn how energy is used for a variety of reasons (a school of engineering or trade school might be such a place)

There are now a bewildering variety of energy and water monitors and controls on the market, with software and systems that can give instantaneous readings of usage, demand charges and other penalties, operating metrics for parameters which are out of bounds, temperature, air flow, and many other measurements. Where only general information is needed, they are installed at the service entrances or on main connection boxes; for specific operating details on manufacturing processes or pumps, they are installed at the load. Communications can be wired, wireless, and even in mesh wireless systems which communicate through whichever operational nodes are closest. Systems can provide monitoring functions only, or they can control systems until they are out of bounds and then alarm the operator. They can be purchased and installed/managed in-house and use cloud-based services, or the hardware and software can be rented or leased. Most electrical sensors can be installed without rewiring, and most gas and water monitors can be installed on the outside of the pipes.

Finally, utilities are becoming visible, trackable, measurable, and predictable. Everything can be tracked.

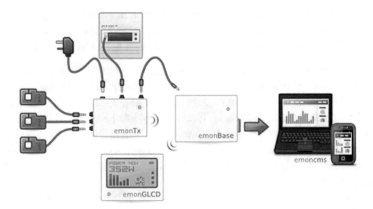

Fig.   9.1 Connection   schematic.   (Cantor   (2017).   http://heatpumps.co.uk/technical/
energy-monitoring/)

For illustrative purposes, some examples of the types of output available from a
few different systems are shown below. Before choosing a system, managers will
want to conduct the internal analyses detailed in the next section so that they can
appropriately specify their needs.

9.7   Examples of Outputs from Different Energy Monitoring Systems (Figs. 9.1, 9.2, 9.3 and 9.4)

Energy monitors are highly adaptable and flexible, providing useful information
on an individual heat pump as in Fig. 9.3, and for multiple buildings as in Fig. 9.5
which shows the demand, temperature, and load segmentation for several Microsoft
buildings.

Without monitoring or control systems, real-time utility use can only be estimated and the reasons for anomalies only guessed at. In addition, the baseline
against which results are measured must be calculated based on these estimates.
That can be done manually by analyzing whole bill data on total savings, calculations of reduced demand or other penalties, information on reduced usage over a
period of time, etc. Managers can then use that information in combination with
other variables in their business to verify results. It is a time-intensive process, however, and managers must be highly motivated to continue the process. Monitors on
the other hand can be magic.

9.8   Incentives and Paying for Projects

For those who decide to invest in energy effectiveness or renewable energy projects,
there are multiple ways to finance the investment. As with purchasing a car, the easiest approach is pay cash. On the other hand, it may also make sense to borrow the

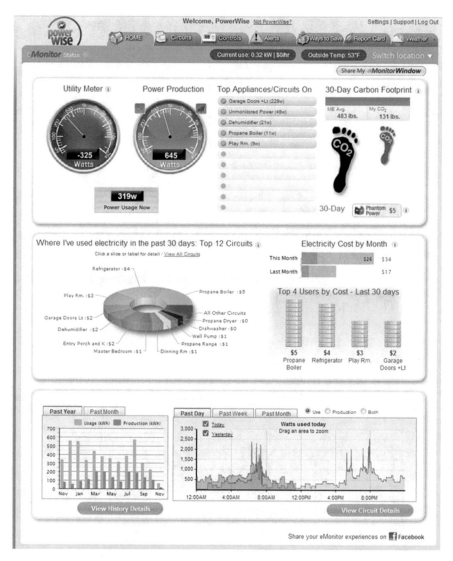

Fig. 9.2 Energy monitoring systems are cost-effective at small businesses and in residences. Example of dashboard. (Powerhouse Dynamics (2017). EnergySage/Used with Permission)

funds, use the utility's incentive funds, issue a bond, or hire an energy service company to do the work required on a performance contract. Not surprisingly, projects designed to increase positive cash flow are attractive to financing institutions.

Often, the amount of loan required can be reduced by incentives and grants offered by different entities for various reasons including:

- Local economic development in both manufacturing and service sectors.
- Reductions in greenhouse gas emissions, pollution, or other undesirable consequences of energy generation and use.

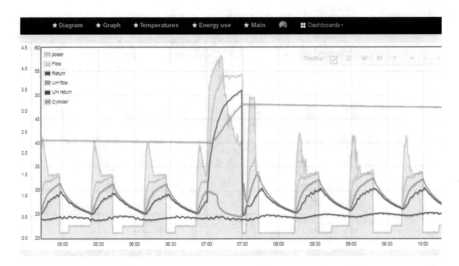

Fig. 9.3 Heat pump energy usage by component. (Cantor (2017). http://heatpumps.co.uk/technical/energy-monitoring/)

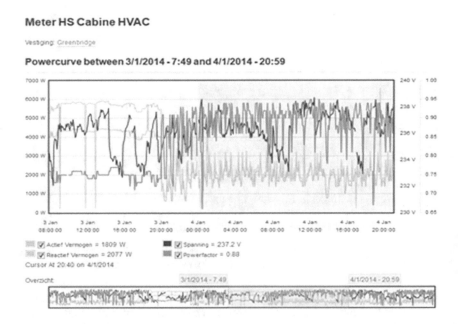

Fig. 9.4 Monthly power curve for HVAC. (Argus Technologies (2014-2017). https://argustech.be/en/energiemonitoring/)

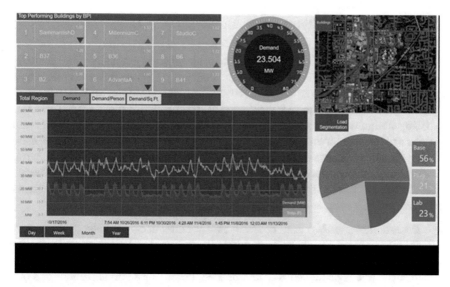

Fig. 9.5 Data analytics example. (Microsoft (2016). https://www.microsoft.com/itshowcase/
Article/Content/845/Data-analytics-and-smart-buildings-increase-comfort-and-energy-efficiency)

- Postponement of additional investments in large-scale generation facilities by the utility.
- Reduced reliance on energy from unstable markets.
- Increase competitiveness of local organizations.
- Achievement of marginally related policy objectives.
- Organizations such as governments, foundations, utilities, and others often develop frameworks and incentives to drive market adoption of various energy efficiency or alternative energy products and procedures, and managers should be aware of and take advantage of these incentives where appropriate.

The energy market has multiple components, the effects of which may be on balance positive or negative, and a market has been developing internationally to drive reductions in the harmful effects. Using Green tags for renewable energy and white tags for energy efficiency, the positive attributes of each kWh of energy provided are separated from the kWh themselves. There is a market in these secondary attributes, sometimes called Renewable Energy Certificates [2] or RECS.[3]

---

[2] California Green Solutions. (n.d.). http://www.californiagreensolutions.com/cgi-bin/gt/tpl.h,content=2506. Accessed 1/10/2017.

[3] Worldwatch Institute. (2016). http://www.worldwatch.org/node/5135. Accessed 1/10/2017.

9.9   Funding Methodologies

In order for an energy or water project to be properly evaluated, the financing mechanism must also be determined. Most energy programs focused on low-cost/ no-cost efforts, operations, and maintenance or human behavior approaches would probably be considered an operating expense, while investments in a new furnace or roof insulation would likely be capitalized. The financial analysis should therefore consider – as with any other project – the trade-off or opportunity cost of using internal funds vs. financing the investment. The analysis should be no different from that for any other operating expense or capital investment anticipated to provide a positive return. Prior to beginning that process, the availability of incentives, grants, rebates, or other sources of outside capital should be investigated.

- *Tax rebates/incentives* may be available for either energy efficiency projects or alternative energy projects from national, state, and/or local governments. The amount received from such payments depends on the business' tax situation, so financial and legal advice is recommended. For nonprofit organizations and others which receive no benefit from tax incentives, it is sometimes possible to sign a power purchase agreement with individuals or groups who are interested in the tax benefits.

  - *Green tags or renewable energy credits* have been created in some areas where the utilities are obligated to have a certain percentage of power generated by renewable energy. The generation itself thus has value to the utility, which pays for the renewable energy attribute, generally as power is produced.
  - *Grants* may, under some circumstances, be available from government entities or others for alternative energy projects. Generally, these grants are short term and have specific requirements; the DSIRE website (dsireusa.org) is likely to have up-to-date information for most areas in the United States.
  - *Energy efficiency programs* are offered by many utilities in the United States and in some other countries. These provide incentives to customers who engage in energy efficiency programs. The programs are paid for by their customers in part, and the utilities support them in order to comply with regulations and to reduce the need for constructing new generation plants.

- Other funding mechanisms include:

  - *Power purchase agreements* are normally used for large renewable energy installations; a contractor or investor covers the cost of an alternative energy project, and the client or host property is charged a reduced rate for the power purchased. Contracts set out the terms of the agreements for both sides.
  - *PACE arrangements* (property-assessed clean energy) move the cost of the renewable energy system to the property tax bill, with the company installing the solar system being paid by the taxing authority.
  - *Carbon fund financing* is available in some countries, where the reduction in carbon – if carefully quantified and certified – has value.

- *Loans* can come from normal sources as well as institutions which specialize in energy efficiency or alternative energy projects.
- *Bonds* can generally be arranged for government borrowers from the same institutions which provide loans for energy projects.
- *Capital or operating leases* can often be arranged at the borrower's option since energy projects are generally multifaceted.
- *Energy service companies*, or ESCOs, install large and often varied energy projects without requiring an investment from the client. The client pays the ESCO for maintenance and operations over a period of time. Savings based upon the performance documented by the ESCO are shared, with the ESCO's risk compensation and profit coming out of their portion of the savings. These are called performance or shared savings contracts.
- *Affiliated groups* may include tenants, funding institutions, suppliers, or others who may be interested in covering part or all of the cost for an alternative energy or energy efficiency improvement. This may be either as a charitable contribution or as a financial transaction with expectation of repayment plus interest.

For a list of current options and explanations in the United States, see www. dsireusa.org, the Database of State Incentives for Renewables and Efficiency produced by the NC Clean Energy Technology Center.

In summary, there are many ways that energy projects can be funded. If they make financial sense, it is highly likely that a banking institution, a government entity, an investment fund, or some other group will be interested in providing the financing. There are also ways to become more energy effective without significant investment. One approach is to grow an energy-effective organization by starting with low-cost/no-cost options, capturing the savings to be dedicated to larger investments and increasing benefit.

References

ACEEE, American Council for an Energy Efficient Economy. (n.d.). Programs. Retrieved from ACEEE.org: http://aceee.org/portal/programs.

Alliance to Save Energy. (2013). *The history of energy productivity.* Washington DC: Alliance to Save Energy. Retrieved from http://www.ase.org/sites/ase.org/files/resources/Media%20 browser/ee_commission_history_report_2-1-13.pdf.

Argus Technologies. (2014–2017). energiemonitoring. Retrieved from argustech.be: https:// argustech.be/en/energiemonitoring/.

California Green Solutions. (n.d.). White Tags, Green Tags, etc. for Renewable Portfolio Standard Programs. Retrieved from www.californiagreensolutions.com: http://www.californiagreensolutions.com/cgi-bin/gt/tpl.h,content=2506.

Cantor, J. (2017). Energy monitoring. Retrieved from heatpumps.co.uk and heatpumps.co.uk: http://heatpumps.co.uk/technical/energy-monitoring/.

Cantore, N. (2014). Factors affecting the adoption of energy efficiency in the manufacturing sector of developing countries. *Energy Efficiency.* https://doi.org/10.1007/s12053-016-9474-3.

Carbon Trust. (n.d.). ctv001_retail.pdf Sector Overview. Retrieved from www.carbontrust.com: https://www.carbontrust.com/media/39228/ctv001_retail.pdf.

cdn.com. (n.d.). Water Demand Forecasting. Retrieved from image.slidesharecdn.com: https://
    image.slidesharecdn.com/waterdemandforecasting-150316010626-conversion-gate01/95/
    water-demand-forecasting-3-638.jpg?cb=1426982084
Chiaroni, D., et al. (2016). Overcoming internal barriers to industrial energy efficiency through
    energy audit; a case study of a large manufacturing company in the home appliances industry.
    Clean Technology Environmental Policy.
Eggink, J. (2007). *Managing energy costs - a behavioral and non-technical approach*. Lilburn:
    Fairmont Press.
Gerarden, T. G. (2015, January). Addressing the Energy-Efficiency Gap Faculty Research Working
    Paper Series RWP15–004. Retrieved from Harvard Kennedy School: https://research.hks.har-
    vard.edu/publications/workingpapers/Index.aspx.
Hansen, S. (n.d.). Making the Business Case for Energy Efficiency. Retrieved from docplayer.
    net: http://docplayer.net/6344792-Making-the-business-case-for-energy-efficiency-shirley-j-
    hansen-ph-d.html.
Laitner, J. A. (2012). *The long-term energy efficiency potential: What the evidence suggests, report
    number E121*. Washington, DC: American Council for an Energy-Efficient Economy.
Leonardo Academy. (n.d.). Free courses and programs on sustainable energy. Retrieved from
    Leonardo Academy: www.leonardo-academy.org.
Microsoft. (2016). Data analytics and smart buildings increase comfort and energy efficiency.
    Retrieved   from   www.microsoft.com: https://www.microsoft.com/itshowcase/Article/
    Content/845/Data-analytics-and-smart-buildings-increase-comfort-and-energy-efficiency.
Mitsubishi Heavy Industries. (n.d.). History of fossil fuel usage since the industrial revolution.
    Retrieved from MHI.com: https://www.mhi-global.com/discover/earth/issue/history/history.
    html.
Mourik, R., et al. (2015). What job is energy efficiency hired to do? A look at the propositions
    and business models selling value instead of energy or effvviciency. Retrieved from IEADSM
    Leonardo Energy: https://www.youtube.com/watch?v=GGLYp_fHrMs.
Powerhouse Dynamics. (2017). Solutions - Screen captures sent via personal communications.
    Retrieved from www.poerhousedynamics.com: https://powerhousedynamics.com/solutions/
    sitesage/.
Public Technology Inc. /US Green Building Council. (1996). *Sustainable building technical man-
    ual*. Washington, DC: Public Technology, Inc.
Thorpe, D. (2013, July 1). Demand Side Response: Revolution in British Energy Policy. Retrieved
    from The Energy Collective.com: http://www.theenergycollective.com/david-k-thorpe/244046/
    demand-side-response-revolution-british-energy-policy.
US Department of Energy Energy Efficiency and Renewable Energy. (2011). A Guide to Energy
    Audits PNNL-20956. Pacific Northwest National Laboratory. Retrieved from http://www.pnnl.
    gov/main/publications/external/technical_reports/pnnl-20956.pdf.
Worldwatch Institute. (2016). Green Tags. Retrieved from www.worldwatch.org: http://www.
    worldwatch.org/node/5135.

Part III
Introducing and Using the Strategic Energy Effectiveness Framework (SEE): Strategy

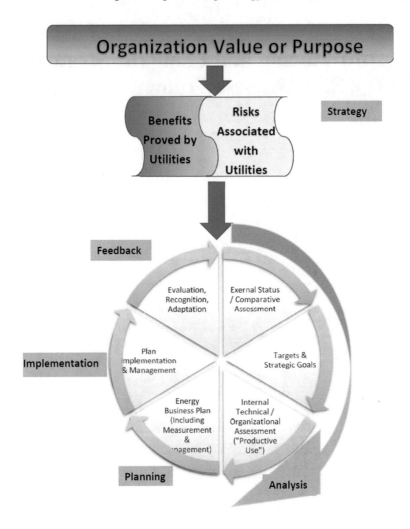

In order to begin using the Framework, it is first important to connect the organization's Purpose and Value to operations and activities. For energy and water, that process also involves understanding how utilities operate so that the risks and benefits of working with those utilities can be internalized. Graphically, Strategy represents the beginning of the process.

STRATEGIC ENERGY EFFECTIVENESS FRAMEWORK

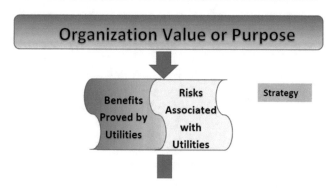

Fig. 1 SEE Framework, Strategy Portion

Chapter 10
Energy and Water in the System and for a Purpose

10.1 General Comments

In order for true energy effectiveness to be achieved there must be a strong connection between purpose and value and the way utilities form part of the fabric of the organization. This requires a paradigm shift so that managers and staff understand energy and water in the context of the organization, and actions are taken in a systems context with utilities seen as controllable costs, or even controllable assets.

Those who work in the alternative energy and energy efficiency arenas enjoy the work partly because it is intellectually challenging, an industry where what works for one client is unlikely to work in the same way for another. Most clients on the other hand would prefer packaged solutions, where they can take a system that worked elsewhere and simply adopt it as is. The Framework bridges that gap by providing a systematic and systemic approach; it is a different way of looking at businesses and the roles energy plays within them, coupled with a path to action and continuous improvement.

Part III builds upon the points made in Part II concerning energy generation, use, and waste. Using the SEE Framework to approach energy issues by drawing upon long-standing business practice, real-time information, and human behavior, managers can strengthen their organizations in both expected and surprising ways.

The first enterprise-wide decision of consequence is the determination of what business the company is in. That single factor drives many operational and energy requirements, which are likely to remain relatively constant. The business type does however determine the relevant comparative models, leading towards a particular set of potential energy effectiveness directions. Whatever form the enterprise takes, significant increases in profitability through energy effectiveness are possible.

© Springer International Publishing AG, part of Springer Nature 2018
S. McCardell, *Energy Effectiveness*, https://doi.org/10.1007/978-3-319-90255-5_10

10.2   Systems Thinking

Organizations, logically, are systems. If an inventory control person is out ill for an unexpectedly long period of time, for instance, and nobody else is assigned his/her duties, inventory issues are likely to plague every part of that organization. If a lighting upgrade is approved, that lighting upgrade should provide the light quality needed for particular detailed tasks while also reducing costs. Manufacturing processes are systems designed to achieve a specific result, as are heating and cooling, inventory, water/wastewater, and finance – indeed, most parts of an organization. In other words, organizations are not only systems in themselves – they are composed OF systems. In such a context, systems thinking works better than linear thinking.

Systems Thinking Example

Imagine a mid-sized office building, with high ceilings and fluorescent tube lighting. On Thursday afternoon, the facilities manager learns that four light fixtures in the office of the most difficult tenant have burned out, and since that tenant works over the weekend, the facilities manager knows that she had better replace the lights on Friday in order to avoid being called in over the weekend. Unfortunately, there are no lights or ballasts in inventory, and in fact the facilities manager is not sure whether the bulbs or the ballasts are the problem, so on her way to work on Friday, she stops and picks up eight bulbs and four ballasts (of exactly the same type as the ones that failed) and installs them. One issue, one problem solved, the tenant stops complaining and the facilities manager is off the hook for the weekend. That is linear thinking.

The same problem could also be approached using systems thinking. Imagine that the facilities manager knows lights represent approximately 30% of the energy use in the building and that the wattage/electricity cost reduction per fixture if the fluorescent tubes were replaced 1 for 1 with LED tubes would be approximately 70%. In addition:

- *Perhaps a neighboring tenant has been complaining of headaches lately, believing them to be caused by the fluorescent lights which have started to flicker and hum.*
- *Perhaps the facilities manager is lucky enough to have one tenant who is an electrician, prepared to give her a discounted rate on the labor to install LEDs because he feels that will help his business.*
- *The facilities manager may also remember that all the fluorescent tubes were installed at the same time, about 5 years ago – and she may have started to see a pattern of failure, which is why she has the names of several LED providers on a sticky note by her computer.*
- *The electrician mentions that the utility has incentives for replacing fluorescents with LEDs, and with his labor discount, the net cost for replacing the old fluorescent fixtures with LEDs is slightly more than the retail cost of the like for like replacement of four fluorescent tubes.*

(continued)

> • *Just to make things interesting, let's imagine that the tenant with a headache gladly takes the facilities manager up on her idea to temporarily use desk lamps so that the difficult tenant can borrow bulbs and ballasts for a couple of weeks.*
>
> *While the difficult tenant is working over the weekend using the borrowed lights and the tenant who suffers from headaches determines that in fact the flickering fluorescent lights were causing the problem, the facilities manager does a financial comparison between replacing current lights with similar ones and replacing all fluorescent tubes with LEDs; the analysis demonstrated that the LED option would provide net savings even in the first year, with an ROI of over 70%.*
>
> *Multiple issues were solved and multiple benefits achieved. The facilities manager created goodwill with the tenant whose headaches were such a bother and the electrician who could showcase his work, reduced long-term costs and improved the building's financial performance for her employer, kept the tenant happy, made the building more attractive, and showed that a systems approach can result in multiple improved outcomes.*

"Systems thinking" was first defined by Jay Forrester at MIT in 1956, as part of what he called system dynamics.[1] The concept is now used in many different fields including biomimicry and agile project management, which is predicated on the notion that circumstances will always change and evolve, so the project or organization must adjust systematically and systemically to those changes.

Within organizations, systems thinking has provided a breakthrough to understanding such a complex world; and as a management approach or discipline, it takes into account relevant linkages and interactions, subsystems, and patterns, analyzing and using leverage points to accomplish change along a number of different vectors rather than just inside one "silo," department, or focus area. "It provides a means of understanding, analysing and talking about the design and construction of the organisation as an integrated, complex composition of many interconnected systems (human and non-human) that need to work together for the whole to function successfully."[2]

In the energy context, the value of systems thinking is that it combines elements of engineering and social science, allowing "people to make their understanding of social systems explicit and improve them in the same way that people can use engineering principles" to accomplish the same thing.[3]

---

[1] Aronson, D. (n.d.). http://www.thinking.net/Systems_Thinking/OverviewSTarticle.pdf. Accessed 1 Oct 2016.

[2] Institute for Systemic Leadership (2017). http://www.systemicleadershipinstitute.org/systemic-leadership/theories/basic-principles-of-systems-thinking-as-applied-to-management-and-leadership-2/. Accessed 1 Oct 2017.

[3] Aronson, D. (n.d.). http://www.thinking.net/Systems_Thinking/OverviewSTarticle.pdf. Accessed 1 Oct 2017.

Systems thinking approaches are designed to mitigate the issues associated with unintended consequences by understanding that those effects are logical consequences and could have been anticipated – as the Figs. 10.1, 10.2 and 10.3 show.

Fig. 10.1 The importance of systems thinking (Art of the Future. http://www.artofthefuture.com/images/gif/STcartoon.gif. Accessed 5 Feb 2017)

Fig. 10.2 Systems thinking and unanticipated consequences (Oxfamblogs. http://oxfamblogs.org/fp2p/wp-content/uploads/2015/10/systems-thinking-fail.jpg. Accessed 5 Feb 2017)

Fig. 10.3 A holistic view (mpr. https://media.licdn.com/mpr/mpr/AAEAAQAAAAAAAAQHAAA AJDM3ZTBiNDUzLWI2ZTYtNGM4My1hYTNhLWVjZTlmYzMyYzg1NA.jpg. Accessed 5 Feb 2017)

10.3   Energy Source, Energy End Use, and Leverage Points

Effective energy management can focus on the energy source, the end-use equipment, and other categories or points of leverage. Cost recapture opportunities can be found outside the organization as well as inside, as shown in Fig. 10.4; only those within the organization, however, are under the control of management.

Reading the bars from left to right, the left-most column shows the inefficiency of the utility's generation process, which is completely outside management's control. In general, off-site generation, transmission, and distribution are extremely inefficient. For some organizations, it may be possible to generate power on-site with alternative energy, and technical improvements in product and service costs are likely to continue driving the installed cost for such systems down.

The billing bar captures the point at which utilities enter the facility, showing what the organization purchases and the multiple types of charges for those purchases.

The distribution bar captures the way charges are assessed by the utility to increase its revenue, combined with the ways and times energy is used.

A natural gas provider, for example, is always obligated to provide natural gas to its customers, even if the utility is running low on natural gas at the end of a long winter heating season. The utility may have to purchase additional high-priced gas on the spot market, or they may decide to depress consumption by increasing prices during periods when demand is high. One way to do that is to provide lower rates to customers who agree to use gas only when it is plentiful. Utilities use many such techniques, and the precise pricing methods are often very difficult to identify from the utility bills.

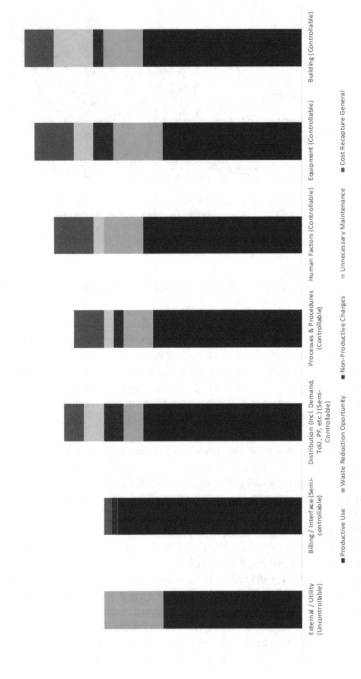

Fig. 10.4 Cost recapture opportunities (Copyright 2016–2018 Current-C Energy Systems, Inc. Reprinted with permission)

The four right-most columns show utility use inside the facility in each of the Four Fields, where the most significant usage and cost reductions can be achieved. Some require investments of time, while others require investments of funds. As the old adage says, "There is no such thing as a free lunch." On the other hand, the benefits for such investments can be multiple and significant, starting with 20–50% potential cost savings.

In every organizational system, there are many different leverage points. When they are addressed as part of a system, many benefits are possible. The sketches below are designed to help readers think in systems terms.

10.4 Using Systems Thinking

Sketch #1

A small fabrication operation is the main employer on a small low-lying island without electricity. There are also no rivers and streams. They manufacture furniture pieces using local wood. Consider the following questions:

- *How do they cut the trees?*
- *How do they transport the logs?*
- *How many hours a day can the operation run?*
- *How do they sell their products?*
- *How do they connect to the outside world?*
- *How do they produce steam to form the furniture?*
- *How do they keep the facility clean?*
- *How do they keep their employees comfortable?*

There are literally hundreds of ways in which even a small operation such as this one depends on the ready availability of power, heating, cooling, and water, generally purchased from a local utility. The fabrication operation would be obliged to use different systems. For example, a solar thermal system might be designed that could produce steam to form the furniture, to run a turbine for power generation, and to cool the building through a heat exchanger.

Think of your organization – what is there that you do that does NOT require power, heating, cooling, or water? Is there anything you can see or think of in your operation that was produced without these building blocks?

Next, consider the impact of change. If you did one thing differently, now would many other systems be impacted in either positive or negative ways?

Sketch #2

A mid-sized company with operations in three different countries decides that their Human Resources Department will go paperless. What effects might that have?

- *How much space currently occupied by paper files would be freed up for other uses?*
- *How would communications between HR departments in each location change?*
- *How much paper and toner would be saved, and what would the cost reduction be?*
- *What would the impact be on security?*
- *Could the number of printers in HR departments be reduced?*
- *Would staff enjoy their jobs more if they spent less time doing paperwork?*

Not all effects would be positive, of course – and for some companies, going paperless would be inappropriate. But for others, there might be positive impacts not only on costs and resource use but also on employee satisfaction. A paperless initiative could also be folded in slowly, if that makes more sense.

Sketch #3

A similar mid-sized company decided to go paperless by department, phasing the process over 4 years. In the first year, they saved $5000, and enthusiastic administrative personnel competed to see who could produce the least paper. The phased process continued more quickly than had been anticipated as negative outcomes were addressed and positive outcomes increased.

References

Arch Tool Box (n.d.). *Energy use intensity.* Retrieved from Archtoolbox.com, Architect's Technical Reference: https://www.archtoolbox.com/sustainability/energy-use-intensity.html

Aronson, D. (n.d.). *Systems thinking overview.* Retrieved from: www.thinking.net: http://www.thinking.net/Systems_Thinking/OverviewSTarticle.pdf

Art of the Future (n.d.). *STcartoon.* Retrieved from Art of the Future: http://www.artofthefuture.com/images/gif/STcartoon.gif

Authenticity Consulting (n.d.). *systemsthinking.pdf.* Filed guide to consulting and organizational development. Retrieved from: www.authenticityconsulting.com: http://managementhelp.org/misc/defn-systemsthinking.pdf

Baker, N. E. (2017). *Institutional change federal energy management.* Retrieved from Energy. gov/eere/femp: https://energy.gov/eere/femp/institutional-change-federal-energy-management

Institute for Systemic Leadership (2017). Basic principles of systems thinking. Retrieved from systemicleadershipinstitute.org: http://www.systemicleadershipinstitute.org/systemic-leadership/theories/basic-principles-of-systems-thinking-as-applied-to-management-and-leadership-2/

mpr(n.d.).*mpr.*Retrievedfrommedia.licdn.com:https://media.licdn.com/mpr/mpr/AAEAAQAAAAAAAAQHAAAAJDM3ZTBiNDUzLWI2ZTYtNGM4My1hYTNhLWVjZTlmYzMyYzg1NA.jpg

Oxfamblogs (n.d.). *Systems thinking fail.* Retrieved from oxfamblogs.org: http://oxfamblogs.org/fp2p/wp-content/uploads/2015/10/systems-thinking-fail.jpg

Chapter 11
Utility Benefits and Risks

11.1 The Utility Business Model

Utilities are outside the influence of any entity which purchases from them. They are either regulated or semi-regulated by various government bodies, and therefore the prices they charge and the investments they make are not always within their own control either. They are required to provide consistent power to customers over the long run, at reasonable rates, and in consideration of those obligations, they are allowed a percentage return on assets and a monopoly or semi-monopoly position.

Utilities are also unusual in several other ways that impact their customers.

- Utility-scale electricity cannot be stored but must be transmitted and used instantly. Where wind turbines are used for generation, some may have to be shut down if there is no demand for the power they produce. If coal is used for generation, production must be calibrated to the anticipated demand and cannot be adjusted easily or quickly; peaker plants (often natural gas) provide the additional capacity to handle load surges.
- Natural gas can be produced and stored, but storage facilities are generally at the generation site rather than distributed throughout the system.
- Water can be and is stored, especially in rural areas which rely on direct pumping of groundwater.
- Most distribution systems are interconnected, except for water in remote areas.
- Utilities produce or distribute only one commodity but at different rates for different customers, and they are committed to providing that product to their customers at a rate which is at least partly regulated. They cover their costs by assessing other fees, some of which are mandated or allowed by regulatory or taxing authorities. The mix is different for each utility.
- In countries and/or cities where brownouts or blackouts are the rule rather than the exception, reliability of service is a more critical consideration than price, whether the utility providing service is a monopoly or not.

© Springer International Publishing AG, part of Springer Nature 2018   127
S. McCardell, *Energy Effectiveness*, https://doi.org/10.1007/978-3-319-90255-5_11

It is not difficult to imagine the challenge faced by utility managers. Utilities in most countries are considered basic needs, which means that governments regulate the companies which provide them. Utility companies are to some extent protected from price competition because of these requirements, but in return they must follow the regulations, meeting demand and market challenges without being allowed to adjust prices quickly. They do this in a variety of ways which are important to understand in order to manage energy costs.

11.2   How Utilities Generate Product and Revenue

Although utility customers tend to view utility bills as providing a single product or service – "Water for August" or "Natural Gas for December" – each monthly invoice includes different categories of charges. Some of them are consistent or fixed, and some are variable.

Consistent charges include base rates which cover the costs for producing and delivering the commodity and service such as the costs of generation facilities, transmission and distribution facilities, water treatment facilities, etc. These charges are approved by the regulatory authorities and must meet tests of reasonableness in each jurisdiction. Some of these base rates are fixed, some are flexible, some are assessed per account, and others are based on usage.

There are also cost adjustments for electricity and natural gas, which take into consideration the fact that market prices for those commodities can be volatile and the generation and transmission companies are promised a reasonable rate of return. Generation costs vary by feedstock, infrastructure cost, transmission costs, and other factors, so each utility may have different cost adjustments depending on the source of their fuels. In any particular bill, different rates may be assessed if the commodity price has fluctuated.

Many commercial/industrial accounts (and residential ones, in some locations) are charged what are essentially penalties for usage during periods of high demand, for high but brief spikes in usage, or for different types of inefficient use, including low power factor or load factor.

Each utility bill may also include other fees assessed for specific purposes and approved by the regulatory authorities, such as charges which fund infrastructure refurbishment, energy efficiency programs, taxes, and others. It is highly likely that these types of charges will continue to increase, because the infrastructure for electricity generation and transmission, natural gas transmission, and water/wastewater treatment and distribution is generally aging and therefore in need of additional maintenance, replacement, or upgrade.

11.3   Transmission and Distribution

The diagrams shown in Figs. 11.1–11.4 provide a generalized picture of the generation, transmission, and distribution systems for each of the most common utilities – natural gas, electricity, and water.

Figure 11.1 shows several sources for natural gas: gas wells, storage facilities, and imports. Other sources can include coalbed methane production, pipelines from neighboring countries or areas, landfill gas, and recaptured methane from a variety of sources. Gas can be stored until needed, but once it leaves the storage facility and passes through the compressor station, it must be used (Fig. 11.1).

Electricity, an energy carrier rather than a source, can be generated from many different fuels. These range from coal to wind, from solar power systems to biofuels, and from hydropower to natural gas. Unlike gas, electricity cannot be stored (unless it is first converted to a different type of energy). Frequently, electricity is transmitted over hundreds or thousands of miles in high-voltage transmission lines and then stepped down by transformers to enter the distribution lines which provide electricity to end-use customers (Fig. 11.2).

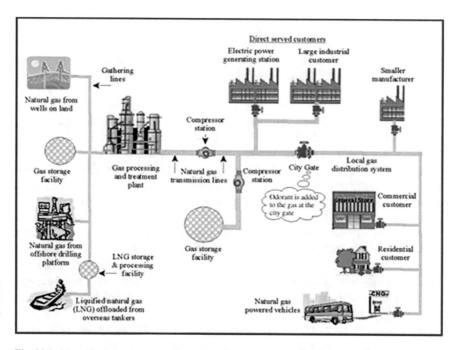

Fig. 11.1   Natural gas distribution system. (Pipeline and Hazardou Materials Safety Administration. https://www.phmsa.dot.gov/portal/site/PHMSA/menuitem.6f23687cf7b00b0f22e4c6962d9c8789 /?vgnextoid=4351fd1a874c6310VgnVCM1000001ecb7898RCRD&vgnextchannel=f7280665b9 1ac010VgnVCM1000008049a8c0RCRD&vgnextfmt=print. Accessed 13 Apr 2017)

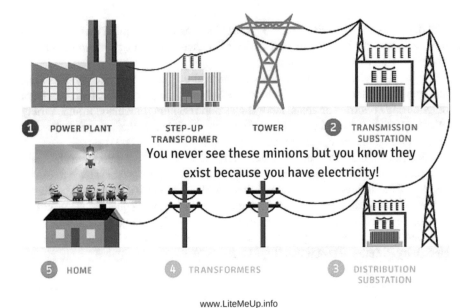

Fig. 11.2 Electricity distribution system. (www.LiteMeUp.info. http://thekoreancouponer.com/wp-content/uploads/2015/02/Energy-home-delivery-graphic.png. Accessed 13 Apr 2017)

Water utilities provide not only the treated/potable water we purchase to use but also the wastewater we dispose of, which is then treated and reused in various ways. There are therefore two different cycles associated with water.

As shown in Fig. 11.3, drinking water comes from a source which can be a well, a stream, an aquifer, or a reservoir. It is treated to remove impurities, bacteria, and organic compounds, stored, and then distributed to end users, generally with the use of pumps.

Treatment of water that has been used goes through multiple stages, and wastewater treatment plants are power intensive with many pumps, aerators, controls, and other systems. Output from wastewater treatment plants is returned to rivers or streams, injected back into the groundwater, reused as gray water for irrigation, or targeted for other nonhuman consumption. The cycle is shown in Fig. 11.4.

Understanding how utilities bill their customers provides important background needed to strategize ways to reduce energy and water costs as well as usage.

Component of Water Supply System

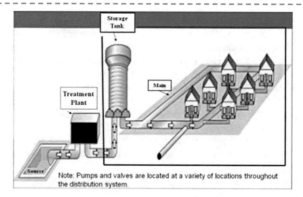

▸ (1). Source   (2). Treatment plant
▸ (2). Storage Tanks/Reservoirs   (3). Water Transmission/distribution

Fig. 11.3 Water distribution system. (cdn.com. https://image.slidesharecdn.com/waterdemand forecasting-150316010626-conversion-gate01/95/water-demand-forecasting-3-638.jpg?cb= 1426982084. Accessed 13 Apr 2017)

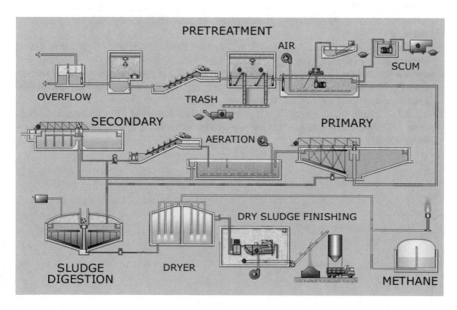

Fig. 11.4 Wastewater treatment and water purification system. (Source: Boundless (2016, May 26). Wastewater and sewage treatment. *Boundless Microbiology Boundless*. Retrieved 13 Apr 2017 from https://www.boundless.com/microbiology/textbooks/boundless-microbiology-textbook/ industrial-microbiology-17/wastewater-treatment-and-water-purification-200/wastewater-and-sewage-treatment-1006-8716/ (Boundless, 2016))

References

Baker, N. E. (2017). *Institutional change federal energy management*. Retrieved from Energy. gov/eere/femp: https://energy.gov/eere/femp/institutional-change-federal-energy-management

Boundless (2016, May 26). *Industrial microbiology/wastewater and sewage treatment*. Retrieved from Boundless.com: https://www.boundless.com/microbiology/textbooks/boundless-microbiology-textbook/industrial-microbiology-17/wastewater-treatment-and-water-purification-200/wastewater-and-sewage-treatment-1006-8716/

cdn.com (n.d.). *Water demand forecasting*. Retrieved from image.slidesharecdn.com: https://image.slidesharecdn.com/waterdemandforecasting-150316010626-conversion-gate01/95/water-demand-forecasting-3-638.jpg?cb=1426982084

Chiaroni, D. E. (2016). *Overcoming internal barriers to industrial energy efficiency through energy audit: A case study of a large manufacturing company in the home appliances industry*. Clean Technology Environmental Policy.

Economist Intelligence Unit. (2012). *Energy efficiency and energy savings: A view from the building sector*. London: The Economist Intelligence Unit LTD.

Gerarden, T. G. (2015, January). *Addressing the energy-efficiency gap faculty research working paper series RWP15-004*. Retrieved from Harvard Kennedy School: https://research.hks. harvard.edu/publications/workingpapers/Index.aspx

Pipeline and Hazardou Materials Safety Administration (n.d.). *Gathering pipelines FAQs*. Retrieved from phmsa.dot.gov: https://www.phmsa.dot.gov/portal/site/PHMSA/menuitem.6f23687cf7b 00b0f22e4c6962d9c8789/?vgnextoid=4351fd1a874c6310VgnVCM1000001ecb7898RCRD& vgnextchannel=f7280665b91ac010VgnVCM1000008049a8c0RCRD&vgnextfmt=print

Powerhouse Dynamics (2017). *Solutions – screen captures sent via personal communications*. Retrieved from www.poerhousedynamics.com: https://powerhousedynamics.com/solutions/ sitesage/

Public Technology Inc./US Green Building Council. (1996). *Sustainable building technical manual*. Washington, DC: Public Technology, Inc.

Thorpe, D. (2013, July 1). *Demand side response: Revolution in British energy policy*. Retrieved from The Energy Collective.com: http://www.theenergycollective.com/david-k-thorpe/244046/ demand-side-response-revolution-british-energy-policy

www.LiteMeUp.info (n.d.). Energy home delivery graphic. Retrieved from thekoreancouponer. com: http://thekoreancouponer.com/wp-content/uploads/2015/02/Energy-home-delivery-graphic.png

Chapter 12
Utility Bill Analysis

12.1 Introductory Comments

Following the description of utilities in Chap. 11, this chapter focuses on the bills customers receive from utilities along with some ideas on how to reduce those bills.

Most utility companies have different rate schedules for residential customers, commercial customers, and large industrial consumers. Although there are some exceptions (e.g., California in the United States and Germany), in general the residential customers pay a higher rate per unit, and the industrial customers pay the lowest – a structure analogous to the quantity discount for ticket purchases to a sporting event. Policy changes to discourage excessive use are reversing that pattern in some places, especially for water in drought-stricken areas which charge less for the first usage block and more subsequent blocks of usage.

Some utilities also assess a penalty for usage during times when they are or project that they may be capacity constrained, such as during late afternoons in summer when office buildings and residences both experience high heat gain, requiring more electrical power than earlier in the day. (For a more detailed description of the specific terms used in this chapter, please see Chap. 2.)

12.2 Electricity

Figures 12.1, 12.2, and 12.3 present different examples of commercial electricity invoices. They show the account number and meter number, the "rate schedule" (Electric Commercial Service/E8T or Tariff 850-Medium General Service), and then different categories of charges. The utility's website generally will have more detail on the components of each Tariff or rate structure, how charges are calculated, and other relevant information, or the customer service department can send that information

© Springer International Publishing AG, part of Springer Nature 2018
S. McCardell, *Energy Effectiveness*, https://doi.org/10.1007/978-3-319-90255-5_12

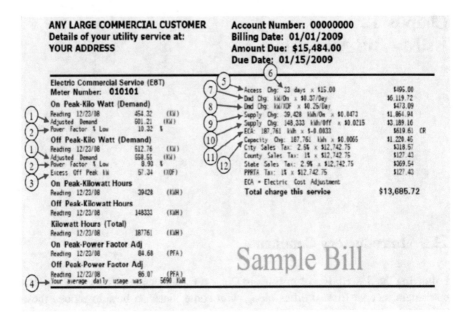

Fig. 12.1 Sample electricity bill #1 – Alpha Energy

out in writing. The rate schedule determines the types of charges that will be assessed and the price. The invoices generally do not show specifics, just the end result.

The invoice in Fig. 12.1 is from an electricity provider called Alpha Energy for the purposes of this chapter. Alpha charges higher rates during high usage times of day (on peak) than during off-peak hours. (The rate schedule would specify when those on-peak/off-peak hours fall.)

Additional charges come from on-peak demand (which is the instantaneous requirement for higher than normal electricity requirements) of 601.21 kW. This translates into a demand charge of $6119.72, and off-peak demand of 658.65 kW translates into a demand charge of $473.09 after being adjusted by the excess off-peak kW. Although the calculations for each charge on the bill are shown, the rate schedule will show more detail.

Some utilities use a ratchet clause, which includes a provision that the highest demand charge in a month will continue to be assessed over a period of time – sometimes as much as 12 months. In other words, the cost consequence of a single action or event that causes a spike in demand cost of $6000 in a month may also be assessed for another 12 months – even if there are no demand charges actually incurred over that period.

Alpha Energy also penalizes the customer for low power factor (a consequence of inefficient use due to inefficient equipment, harmonics, or other electrical issues). Usage charges, shown in kWh or kilowatt-hours, are differentiated by time of day as well, so that the Supply Charges for on-peak use are higher per kWh than for off peak, and the Capacity Charge is based on the total kWh. The other charges or credits include an electric cost adjustment and various taxes and fees which would have been approved by the public regulators.

Energy Charges

Basic Charge	40.00
System Usage Charge (104700.000 kWh x 0.84000¢)	879.48
Off-Peak Usage of 41100.000 kWh x 4.95600¢	2,036.92
On-Peak Usage of 63600.000 kWh x 6.45600¢	4,106.02
Off-Peak Demand 198.000 KW x $0.0000000	0.00
On-Peak Demand 230.000 KW x $2.3300000	535.90
Reactive Demand Charge 9.000 x $0.5000000 Billed KVAR	4.50
(Reactive Demand of 100.500 Actual KVAR)	
Transmission Charge	
230.000 KW x 75.00000¢	172.50
Distribution Charge	
Facility Capacity 30.000 KW x $2.7400000	82.20
Facility Capacity 250.000 KW x $2.6400000	660.00
	8,517.52

Adjustments (see sidebar for more information)

105 Regulatory Adjustments (104700.000 kWh x - 0.02500¢)	26.18	c
109 Energy Efficiency Funding Adj (104700.000 kWh x 0.23700¢)	248.14	
110 Energy Efficiency Customer Svc (104700.000 kWh x 0.00500¢)	5.24	
122 Renewable Resource Adjustment (104700.000 kWh x - 0.01300¢)	13.61	c
123 Decoupling Adjustment (104700.000 kWh x - 0.00900¢)	9.42	c
137 Solar Payment Option Cost Recov (104700.000 kWh x 0.04300¢)	45.02	
145 Boardman Decommissioning Adj (104700.000 kWh x 0.03700¢)	38.74	
	287.93	

Taxes and Fees

Low Income Assistance	52.35
Public Purpose Charge (3%)	256.73
	309.08

Current Charges	**9,114.53**

Fig. 12.2 Sample electricity bill #2 – Beta Energy

Electricity provider Beta Energy uses a rate structure that contains similar elements but is more clearly laid out, and they too are quite transparent, providing calculations for each category instead of forcing the customer to refer to the rate schedule to see the specific costs. Energy Charges are separated into a commodity purchase or usage portion, followed by the demand charges and then the transmission and distribution charges. Beta's invoice also highlights the common step function approach to the utility charged, with the first 30 kW assessed a higher distribution charge than the remaining 250 kW.

Charges for transmission and distribution cover the cost to deliver the power to the customer. The Basic Charge is a monthly fee set based on the rate structure, and the adjustments are to cover costs as approved by the Public Regulation Commission.

Of particular note are the "energy efficiency funding" and "energy efficiency customer service" charges. Most utilities which have incentives for energy efficiency projects installed by their customers assess all customers for the cost of the program, which means that those who do not take advantage of the utility's programs are subsidizing those who do.

Gamma Energy provides very little information in its monthly bill, so the customer will need to refer to the rate schedule for Tariff 850-Medium General Service for specific information.

Rate Tariff: Medium General Service-850		Page 1 of 2
Account Number	**Total Amount Due**	**Due Date**
	$5,653.78	Aug 8, 2016
Meter Number	**Cycle-Route**	**Bill Date**
		Jul 15, 2016

Previous Charges:

Total Amount Due At Last Billing	$	5,758.57
Payment 07/07/16 - Thank You		-5,758.57
Previous Balance Due	$	.00*

Current   Charges:

Tariff 850 -Medium General Service 07/11/16
Service Delivery Identifier:

Transmission Service	$	515.12
Distribution Service		1,541.66
Customer Charge		9.04
Retail Stability Rider		173.15
Deferred Asset Phase-In Rider		46.94
Current Electric Charges Due	$	2,285.91*

Current   Energy Charges

Service Delivery Identifier Number :

Electric Energy Charge (51080KWH @ $0.063463)	$	3,241.69
PJM Capacity Performance (28 Days @ $4.5063920/Day)		126.18
Current Supplier Charges	$	3,367.87*

Total Amount Due	**$5,653.78**
Charges make up the "Total Amount Due".	
Due Aug 8, Pay $5,768.08 After This Date	

Fig. 12.3 Sample electricity bill #3 – Gamma Energy

Where the electricity utility is deregulated, the commodity purchase and service delivery may be provided by different suppliers, and in such a case, one of the utilities (usually the one providing delivery) will likely invoice for both sections. Electricity contracts are also available in deregulated states, and those contracts will provide details on the charges, including whether they are on a fixed or variable rate.

12.3   Understanding and Using Real-Time Data

Understanding what is in the monthly electricity bills is an important first step because the actions necessary to reduce cost, usage, or both may rely in large part on the relationship between electricity usage and rate schedule.

Using Data in Real Time: Example

For example, imagine a small commercial office in an historic building in Florida. The air handling system struggles to keep up with the building's cooling needs, especially on humid summer afternoons. The compressors and fans operate at maximum speed, finally overcoming the heat issue which allows them to slow down, reducing the electrical load. Towards the end of the day, cooling needs increase again. This high demand for air conditioning creates a peak in demand for which the utility charges a penalty, because that is the period when most of their customers have high-usage profiles.

The office building cannot turn their air handling equipment off, but they can take advantage of the utility's demand response programs to adjust their use either automatically or manually, avoiding the penalty period. The building manager's other options would be to decrease cooling needs by improving building insulation or to increase the efficiency of the air conditioning equipment; either option would improve energy efficiency at the facility.

The potential effects of both options are shown in Fig. 12.4. Energy efficiency drops the entire use profile, so that in this case the commercial office building would reduce usage and cost throughout the day. Using demand control, the peak usage is decreased, and over that period the air handling equipment is controlled to avoid the higher rates. Electricity use is only reduced for that time period, but the attendant cost reduction may be over 50%, depending on the demand penalty assessed by the utility.

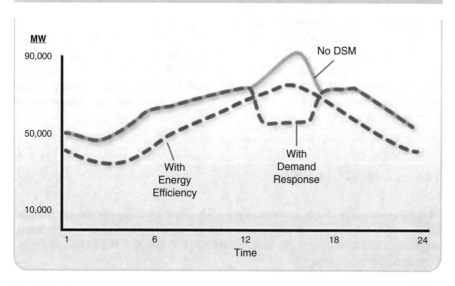

Fig. 12.4 Comparison between demand control and energy efficiency. (Thorpe (2013). http:// www.theenergycollective.com/david-k-thorpe/244046/demand-side-response-revolution-british-energy-policy. Accessed 13 Aug 2017.)

Fig. 12.5 Graph of 15-min data from utility

Another usage pattern that can be adjusted for significant savings is demonstrated in Fig. 12.5, produced from 15-min usage data provided by the utility. The graph clearly shows that the control system schedule is the same every day. The equipment turns on, increases its energy draw until a set point is reached, and then turns off, only to start the cycle over again. This pattern would be typical of HVAC equipment or manufacturing/process equipment, and by adjusting the set points to avoid the peaks and the valleys, savings are possible and desirable if the adjusted schedule has no negative operational impact (Fig. 12.5).

Real-time data enables managers to see patterns, problems, maintenance issues, process and procedure challenges – as well as useful and effective patterns – at the time they are happening, and in time to prevent, alleviate, or minimize electrical issues which have the potential to be either costly or disastrous. Both monitoring systems and control systems, used carefully, can be invaluable to managers in all sizes of businesses.

12.4   Natural Gas Bills

Unit measurements shown on natural gas bills differ by utility. Some use cost per therm, while others use cost per volume such as cubic foot or Ccf. In simple terms, assuming a constant natural gas heat content of 1037 Btu/Ccf, the following definitions and conversions apply:

Definitions:

> Btu – British thermal units. The heat required to raise the temperature of 1 pound of water by 1 degree Fahrenheit

Ccf – 100 cubic feet
Mcf – 1000 cubic feet
MMBtu – 1,000,000 British thermal units
Therm – 100,000 Btu or 0.10 MMBtu
$ per Ccf divided by 1.037 equals $ per therm

Conversions:

$ per therm multiplied by 1.037 equals $ per Ccf
$ per Mcf divided by 1.037 equals $ per MMBtu
$ per Mcf divided by 10.37 equals $ per therm
$ per MMBtu multiplied by 1.037 equals $ per Mcf
$ per therm multiplied by 10.37 equals $ per Mcf

Natural gas can be stored, so natural gas suppliers are not obligated to be prepared, at any hour and without notice, to deliver quantities of as-yet-un-generated
natural gas to customers. This provides gas suppliers with some flexibility balanced
by the fact that natural gas prices are extremely volatile. Fees assessed for natural
gas production and delivery have fewer components than those for electricity, but
they do provide a methodology for adjusting commodity prices during the month if
the pricing changes.

As shown in Fig. 12.6, Alpha Gas provides natural gas on the Large Volume
Service tariff to a customer; that rate is shown on the bill. There is a service charge
assessed per account, and distribution charges are lower per therm for higher levels
of usage. Alpha Gas apparently has much greater demand for its gas in the winter,
because the demand and balancing charge categories penalize wintertime use.
Societal Benefits are a fee approved by the Public Regulation Commission for this
utility to provide local services. Alpha Gas is in a state which is unregulated for gas,
so the commodity currently supplied by Alpha could be sourced from another sup-
plier, even though that gas would be delivered on-site by Alpha.

Beta Gas, the second example, assesses an access charge approved by the
Public Regulation Commission to cover the cost of maintaining distribution systems and providing customer services, as shown in Fig. 12.7. The GCA or gas cost
adjustment reflects changes in commodity or delivery costs, and the "Capacity
Charge" reflects the charge for "transportation" or delivery. As always, there are
taxes and fees.

Gamma Gas, shown in Fig. 12.8, has a simple bill for this account on Rate 03 C,
Commercial. There is a monthly service fee, a charge per therm, and a Public
Purpose Charge. This bill also clearly shows the billing factor for the meter, which
is the number by which the units on the meter are multiplied to show actual therms.
The small graph of usage (often be found on both electricity and natural gas bills)
clearly shows that this customer uses natural gas almost exclusively for heat in
November and December.

Charges				Rate - LVG	
Delivery					1
Service Charge				$91.89	2
Distribution charge					
First	1000.000 therms	x	$0.0668400	66.84	3
Next	1316.286 therms	x	$0.0440400	57.97	4
Demand	99.707 therms	x	$3.5090082	349.88	5
Balancing charge	2205.920 therms	x	$0.09395540	211.67	
Societal Benefits	2316.286 therms	x	$0.038790050	89.85	
Total Delivery				**$868.10**	6
Supply					
BGSS Commodity	2316.286 therms	x	$0.966873	2239.55	7
Total Supply				**$2239.55**	
Total gas charges				**$3107.65**	8

Usage	Meter 123456789	
Actual Reading Feb 1	17873	
Actual Reading Jan 1	15652	
Difference	2220	
Conversion to CCF	x 1.0120	(CCF = One hundred cubic feet)
CCF Total	2246.640	
Conversion to therms	x 1.031	
Total therms	2316.286	

8

1. LVG is a Large Volume Service tariff. Billing for residential and small commercial customers will often be simpler but typically includes some of the same charges.

2. This is the fixed monthly service charge for the Large Volume Service tariff.

3. Note that these charges, based on **usage**, which are intended to pay for the cost of delivering gas, are lower after the first 1,000 therms. This reflects the need to cover the fixed costs of operating a distribution network.

4. This charge is based on **demand**, in this case the average daily **usage** in therms for the wintertime month with maximum consumption. For this bill the demand rate is approximately $3.51 per therm.

5. The balancing charge is an additional adjustment that accounts for the imbalance between summer and winter gas use. In this case the amount is calculated based on the extent to which the average daily use in winter months exceeds average daily use during the summer. The relatively large amount in this bill reflects the fact that almost all of this customer's use is for heating.

6. For this bill the delivery charges are 37.5¢ per therm ($868.10 divided by 2216.286 therms).

7. BGSS is Basic Gas Supply Service from the utility. Customers can choose to purchase gas from alternate suppliers, in which case the charges for the actual gas used may appear on a separate bill.

8. Dividing total charges by the number of therms used gives the cost per unit (in this case $1.34 per therm).

Fig. 12.6 Sample natural gas bill #1 – Alpha Gas

Gas Commercial Service (G1C)

Access Chg (G1CS): 20 days x $0.393	$7.86
Access Chg (G1CS): 72 CCF x $0.1645	$11.84
Gas Cost (G1CS): 72 CCF x $0.6034	$43.44
GCA (G1CS): 72 CCF x $-0.246	$17.71 CR
City Sales Tax: 2.5% x $45.43	$1.14
County Sales Tax: 1.23% x $45.43	$0.56
State Sales Tax: 2.9% x $45.43	$1.32
PPRTA Tax: 1% x $45.43	$0.45
Access Chg (G1CS): 9 days x $0.393	$3.54
Access Chg (G1CS): 32 CCF x $0.1645	$5.26
GCA (G1CS): 32 CCF x $0.01726	$0.55
Capacity Chg (G1CS): 32 CCF x $0.1073	$3.43
City Sales Tax: 2.5% x $12.78	$0.32
County Sales Tax: 1.23% x $12.78	$0.16
State Sales Tax: 2.9% x $12.78	$0.37
PPRTA Tax: 1% x $12.78	$0.13
GCA = Gas Cost Adjustment	

Fig. 12.7 Sample natural gas bill #2 – Beta Gas

USAGE SUMMARY

Meter   461958

Actual Read	6901	12/31/15
Actual Read –	5703	11/25/15
Units of gas	1198	36 days
Billing Factor x	1.216	
Therms	1456.8	

NW NATURAL
ACCOUNT SUMMARY

Previous Balance	$784.59
Payment(s) Received	784.59CF
Subtotal Balance Forward	$.00

Current Charges

Monthly Service Charge Rate 03C	15.00
Gas Usage 1456.8 @ .88920	1,295.39
Public Purpose Charge	32.63
Subtotal Current Charges $1,343.02	
New Balance	**$1,343.02**

A late charge of 1.8% with a minimum of $3.00 may be assessed on past due balance of $50.00 or more.
You are participating in the WARM program, which adjusts bills if weather between Dec. and mid-May is colder or warmer than normal.
Thank you for your payment.

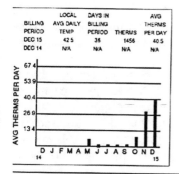

BILLING PERIOD	LOCAL AVG DAILY TEMP	DAYS IN BILLING PERIOD	THERMS	AVG THERMS PER DAY
DEC 15	42 5	36	1456	40 5
DEC 14	N/A	N/A	N/A	N/A

Under normal operating conditions, natural gas burns cleanly. But if the appliance has a mechanical problem, it could create a hazard. If you or other members of the household feel out of breath, dizzy, nauseous, and have headaches, you could be suffering from carbon monoxide poisoning - leave your home, then call 911.

Fig. 12.8 Sample natural gas bill #3 – Gamma Gas

12.5   Water, Sewer, and Stormwater

Water and sewer bills vary even more than electrical and natural gas bills, because the services may be provided by municipalities which are structured-differently from utilities. They can also be combined, where one entity provides both water and sewer services.

As drought has affected many areas of the United States and other countries, water charge tiers in some regions have been inverted from the previous norm to provide incentives for water conservation. The first tier of usage is at a reasonably low rate, but with increasing usage, the cost per gallon increases.

The explanations below are from the water utility in Columbia, MO, and are used with permission. They present an exceptionally complete and well-done description of how water utilities charge for the services they provide, in this case to a residential customer (water, sewer, and stormwater as well as solid waste). And even though the Columbia description is exceptionally well done, it is important to remember that there is a great deal of information on utility websites and available from customer service representatives. Managers should make it a practice to get to know the customer service people and to ask their advice whenever it seems appropriate – they are a valuable but underutilized resource! (Fig. 12.9).

As a further note on utility representatives, they can assist by ensuring that businesses are on the correct rate schedules and by providing information about any relevant utility incentives for energy projects.

It is important for managers to remember that utilities are responsible to assign customers to the appropriate rate schedules when they sign up, but that rate schedule may no longer be the correct one. Utility rules may charge, and organizations may expand or contract. Additional services might be required and/or set up without reference to what is already in place. A company providing services might decide to begin manufacturing or the reverse. All of these changes may impact the rate schedule, and the customer is responsible for identifying mistakes and inaccuracies in current billings. Periodic discussions with utility customer service personnel are therefore important. Utility incentive programs are generally described on websites or flyers received in the mail, but again customer service representatives who are familiar with the programs can help managers determine which, if any, of those incentive programs might be relevant for them.

With water utilities, customer service personnel can also help determine whether or not there might be potential sewer cost reductions. In some jurisdictions, the sewer charges are based on calculated average percentages of potable water discharged into the sewer; in other jurisdictions, it is assumed that 100% of the potable water purchased is then sent back to the sewer. Management may be able to work with consultants or the utility customer service representatives to show that this is not the case and a reduction is due.

Water Meter Charge				$8.30
(12) Water Consumption Tier 1		2.00	$2.79	$5.58
Water Consumption Tier 2		8.00	$3.91	$31.28
(10) Water - PILOT Fee				$3.52
State Regulatory Fee - Water (13)				$0.13
(14) Fire Flow Charge				$1.55
Water Tax (11)				$4.00
Water Total				**$54.36**

City of Columbia, Missouri
Utilites Department   Billing Questions
701 E. Broadway   (573) 874-7380
P.O. Box 1676   ucs@como.gov
Columbia, MO 65205
*Pay Bill (573) 874-7694
*Online - www.como.gov
*Convenience fee applies

Utility after-hours emergency (573) 874-2555

Customer - Account #: 00000000-0000000
Name: Jane Doe,
John Doe
Service Address: 12345 First Street

NEW CHARGES DETAIL

	Usage	Rate	Charge		
Electric Customer Base Charge			$15.60	Donation-Youth Rec Services	$1.00
Electric Consumption Tier 1	300.00	$0.0752	$22.56	**Adjustment Total**	**$8.00**
(9) Electric Consumption Tier 2	450.00	$0.0980	$44.10		
Electric Consumption Tier 3	1,250.00	$0.1336	$167.00		
Electric Consumption Tier 4	1,500.00	$0.1445	$216.75		
(10) Electric - PILOT Fee			$35.08		
Electric Tax (11)			$39.96		
Electric Total			**$541.05**		
Water Meter Charge			$8.30		
(12) Water Consumption Tier 1	2.00	$2.79	$5.58		
Water Consumption Tier 2	8.00	$3.91	$31.28		
(10) Water - PILOT Fee			$3.52		
State Regulatory Fee - Water (13)			$0.13		
(14) Fire Flow Charge			$1.55		
Water Tax (11)			$4.00		
Water Total			**$54.36**		
(15) Backflow Charge			$2.00		
Backflow - PILOT Fee (10)			$0.15		
(11) Backflow Tax			$0.17		
Backflow Total			**$2.32**		
(16) Sewer Base Charge			$11.01		
State Sanitary Sewer Permit Fee (17)			$0.04		
(18) Sewer Flow Charge	8.33	$2.27	$18.91		
Sewer Total			**$29.96**		
(19) Storm Water Charge - Residential			$1.44		
Storm Water Total			**$1.44**		
(20) Solid Waste Charges Residential			$15.42		
Solid Waste Total			**$15.42**		
Donation-Youth Dental Care			$1.00		
Donation-CASH			$1.00		
(21) Donation-HELP			$1.00		
Donation-Public Art			$1.00		
Donation-Beautification			$1.00		
Donation-Fire Dept			$1.00		
Donation-Police Dept			$1.00		

Your bill explained...
• **PILOT (Payment-In-Lieu-Of-Tax):** Equivalent to Gross Receipts Tax. Required by city charter.
• **C.A.S.H. (Citizens Assisting Seniors and Handicapped):** Specifically assists low income elderly and low income handicapped citizens who have exhausted other resources.
• **H.E.L.P. (Heat Energy and Light Program):** Provides aid to low income families with children in one-time emergency situations.
• **Share The Light:** Provides funding that are above and beyond basic services of city government.
• **Miscellaneous charge (other charges that may appear on your bill):** Service deposits, Service charge, Same day service, Meter test, Cut-off charges or Trip out, Billing adjustment, Returned check and service charge, Transferred account balance, Extra dumpster service, Major appliance pick-up, DNR fee, Other.

Fig. 12.9 Sample water and sewer bill (with explanations). (City of Columbia, MO. https://www.como.gov/WaterandLight/Connections/UnderstandingYourUtilityBill.php. Accessed 13 Apr 2017. Used with Permission.)

Utility after-hours emergency: (573) 875-2555

Notes about your service will appear here.

Billing Date	8/17/2016
Previous Balance	$598.98
Payment(s)	($598.98)
BALANCE FORWARD	$0.00
NEW CHARGES are due 9/6/2016	$652.55
TOTAL AMOUNT DUE	**$652.55**

1.5% late fee applies if not paid by due date

When you provide a check as payment, you authorize us to either use the information from your check to make a one-time electronic funds transfer from your account or to process the payment as a check transaction.

Service	Meter	Previous Date	Previous Read	Current Date	Current Read	Meter Multiplier	Usage	Units	Days	Meter Size
Electric	43810L	07/13/16	44538	08/15/16	48038	1	3500	KWH	33	
Water	37863W	07/13/16	1258	08/15/16	1268	1	10	CCF	33	5/8 INCH

3500

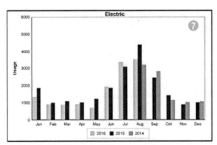

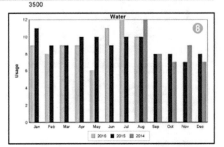

Explanation of billing details

The contact information for the Columbia Utilities billing office is located at the top of the first page of your utility bill. You can pay your bill online, by phone, at the drive-thru or drop box behind City Hall or at our office located on the first floor of City Hall. If you are paying your bill on the due date, please use the drive-thru or pay at the cashiers' office. Convenience fees do apply to payments made by phone or through the City of Columbia's website.

Billing Office: 573-874-7380 or ucs@como.gov

Front page: overview of electric and water usage

The current meter reading (2) is subtracted from the previous meter reading (1) to calculate the usage (3) for the billing period. (4) The days in a billing cycle can vary according to the meter reading cycle. (5) Electric usage is measured in kilowatt hours (kWh). A residential electric customer uses on average 805 kWh a month. (6) Water is measured in Hundred Cubic Feet or CCF. There are approximately 748 gallons per 1 CCF. A residential water customer uses an average 6 hundred cubic feet (CCF) each month.

The weather can make a huge difference in your electric usage. Leaks and irrigation can run up your water bill.

See Columbia Water & Light's efficiency website (ColumbiaPowerPartners.com) to get free tips and incentives to be more efficient.

Fig. 12.9 (continued)

(7) The previous 24 months of your electric usage, measured in kWh, is shown on a graph. You can use this information to follow how weather patterns, energy efficiency upgrades and your habits impact your bill.

(8) The previous 24 months of your water usage, measured in Cubic Hundred Feet (CCF), is shown on a graph. Tracking water usage will help you determine if there are water leaks and/or how much water you are using on irrigation in the summer months.

Back page: details on charges

(9) Electricity is charged on a tiered system, which is now itemized. When a customer's usage hits a certain threshold (300 kWh is the first one) they are bumped into a higher-rate tier. The tier system is not new; only the itemization of it on the bills is new. Tiered rates encourage efficiency and help recover costs associated with the infrastructure needed to supply higher consumption amounts, particularly during the summer months. There are different electric rates for summer and for winter. The new bill format shows the tiers used to figure your bill and the rate for each.

Large commercial customers also pay an electric demand fee because demanding a large amount of electricity all at once goes into the cost to provide electric service.

Taxes

(10) By City Charter, the water and electric utilities are required to collect Payment-in-Lieu-of-Tax (PILOT), an amount equal to the taxes that would be charged for utilities if the utility were privately owned. These funds are transferred to the City of Columbia General Fund and are used for police, fire and other services provided by the General Fund.

(11) Sales tax is applied according to the utility billing amount for electricity and water.

(12) During the summer months, water is charged on a tiered system, which is now itemized. Water usage is measured in hundred cubic feet (CCF). There are approximately 748 gallons per one CCF.

(13) The State Regulatory Fee is a set charge for water testing by the Department of Natural Resources.

(14) The Fire Flow Charge is an additional charge to cover the cost of providing water for public fire protection. The charge is based on meter size.

(15) If you have an irrigation system, you are required to have a backflow protection device. Without one, a drop in the city's water pressure could cause water in your irrigation system to be siphoned into the city's drinking water system. Most commercial and industrial customers are required to have backflow devices in specific locations. State regulations require the water utility to maintain records on all backflow devices in the system and report information annually. The fee is collected to pay for the administrative costs of tracking and reporting these devices.

Fig. 12.9 (continued)

(16) Sewer is a non-metered service that is billed every 30 days or twelve times a year. There is a base charge each billing period to be connected to the system.

(17) State Sanitary Sewer Permit Fees are collected to cover the costs of the State of Missouri charges.

(18) The volume rate (flow charge) you pay is based upon your water consumption or information provided by other water utilities, if you are not a Columbia water customer. Monthly charges for residential customers are based on the average monthly billing of water usage from November through March. For commercial and master meter customers, charges are based on the water used each month.

(19) Storm water is a non-metered service that is billed every 30 days or twelve times a year. The charges are based on the square footage of buildings or other impervious surfaces where rain water cannot be absorbed by the ground. Some commercial customers will see a zero charge as information that the owner of the property is paying the storm water fee.

(20) Trash or solid waste is a non-metered service that is billed every 30 days or twelve times a year. Charges for refuse depend on the type of solid waste pick up the customer receives. For example, there are different rates for bag pick up versus facilities with dumpsters.

(21) Miscellaneous charges you may see on your bill include: service deposits, voluntary charitable contributions (CASH, HELP and Share the Light), major appliance pick-up, etc.

Utility rates

Money collected by the utilities is kept separate from other city revenue. Each utility establishes the utility rate by the cost to serve each customer classification. Most of the utilities have a base charge for being connected to the system, regardless of what you use. For example, meter reading and billing costs are paid through the base charge.

See Chapter 12, 22 and 27 of the City Code of Ordinances for more detailed information on exact charges. The Columbia City Council sets the rates for the utilities.

Stormwater Utility – Chapter 12A

Solid Waste Utility – Chapter 22

Sewer Utility – Chapter 22

Electric Utility

Water Utility

Fig. 12.9 (continued)

References

Federal Energy Regulatory Commission (2015, November). *Energy primer: A handbook of energy market basics.* Retrieved from www.ferc.gov: https://www.ferc.gov/market-oversight/guide/energy-primer.pdf.

International Energy Agency (2015). *Capturing the multiple benefits of energy efficiency.html.* Retrieved from www.iea.org (Jewell, 2014).

Mourik, R. E. (2015). *What job is Energy Efficiency hired to do? A look at the propositions and business models selling value instead of energy or efficiency.* Retrieved from IEADSM Leonardo Energy: https://www.youtube.com/watch?v=GGLYp_fHrMs.

Thorpe, D. (2013, July 1). *Demand side response: Revolution in British energy policy.* Retrieved from The Energy Collective.com: http://www.theenergycollective.com/david-k-thorpe/244046/demand-side-response-revolution-british-energy-policy.

Part IV
Introducing and Using the Strategic Energy Effectiveness Framework (SEE): Analysis

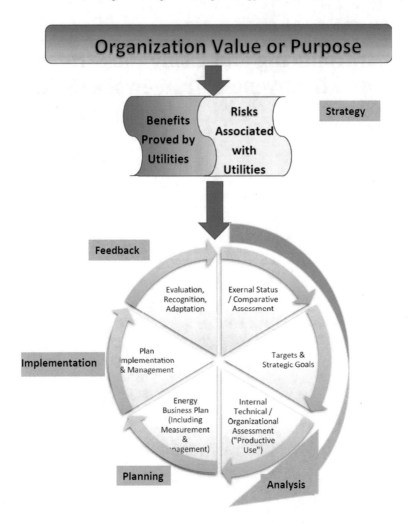

The Analysis stage of the SEE Framework is the most involved, at first. The three stages of analysis are described in Part B, in order.

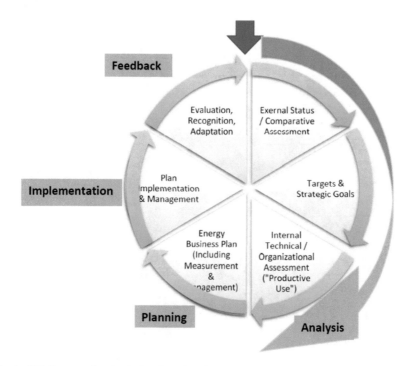

Fig. 1 SEE Framework – Analysis. (Copyright 2016–2018 Current-C Energy Systems, Inc. Used with Permission)

Chapter 13
External Assessment

13.1 Purpose of the External Assessment

When performing a business analysis, it is important to compare the subject business
to others in the community or in the same industry. This benchmarking or relative
evaluation prevents working in a vacuum where there are no clear measures of what
is good and what is unacceptable. For energy projects a similar analysis should be
undertaken, the only difference being that results showing an enterprise to be less
effective than a neighbor are not considered negative; they only present more scope
for improvement and cost savings.

There are two parts to this first piece of the assessment:

- Determine the utility charges in the subject facility.
- Compare that summary to others, and set both a baseline and a tentative target.

Two types of comparisons should be made. The first relates to the financial
statement, and the second relates to the utility bill and utility usage.

13.2 Online Analysis Tools

With an understanding of utility bills from the previous chapter, the next task is to
get a picture of the facility's energy use/waste pattern. Data can be drawn from an
energy monitoring/control system or from the utility bills, but as discussed the data
from automated systems are much more detailed and flexible.

If that data comes from utility bills, the most common approach would be to use
an online tool to input data and then use the tool's calculations and internal database
for comparisons. There are many of these (some of which are offered as part of a
consulting service contract), but since the key for comparisons is to use a tool with
a relevant and robust comparative dataset, there are many free tools available. The
following are some of the ones most commonly used.

© Springer International Publishing AG, part of Springer Nature 2018   153
S. McCardell, *Energy Effectiveness*, https://doi.org/10.1007/978-3-319-90255-5_13

- Green Button Data provides utility usage data in a format that can be easily transferred into other tools to analyze electricity, natural gas, and water use in the United States. An industry-led initiative information is available at http://www. greenbuttondata.org/.
- OpenEI, at http://en.openei.org/wiki/Main_Page, is developed and maintained by the National Renewable Energy Laboratory in conjunction with several other international associations. The site lists many topics; search under "data" for the comparative information desired.
- Energy Star hosts one of the most commonly used tools in the industry for electricity, natural gas (or other heating fuels), and water; waste and materials can also be tracked. Portfolio Manager is the signature Energy Star program for comparisons and for Energy Star certification https://www.energystar.gov/ buildings/facility-owners-and-managers/existing-buildings/use-portfolio-manager, and there are hundreds of pages of tools and resources, including some for particular market segments such as auto dealers, home-based businesses, houses of worship, industrial facilities, and others.
- RETScreen, an energy efficiency and alternative energy project analysis tool developed by Natural Resources Canada, is in frequent use across much of the world and is both flexible and comprehensive. Information can be found at http:// www.energyplan.eu/othertools/allscales/retscreen/ as well as at http://www. nrcan.gc.ca/energy/software-tools/7465.
- The Energy Efficiency Project Resource Center, an international effort, provides economic and financial analysis of energy efficiency projects as well as an enormous body of information on related topics. https://energypedia.info/wiki/ Energy_Efficiency_Project_Resource_Center
- The Commercial Buildings Energy Consumption Survey, or CBECS, is produced by the US Energy Information Administration. The database is both comprehensive and complete and an excellent comparative resource for commercial buildings https://www.energystar.gov/buildings/facility-owners-and-managers/existing-buildings/use-portfolio-manager. It is also incorporated in Energy Star Portfolio Manager.
- In the United Kingdom, https://www.ukpower.co.uk/tools is designed to analyze alternative energy suppliers but also provides calculators to examine specific scenarios.

Some utilities have their own online tools as well, and since utility rates vary, they may provide more locally appropriate comparisons. Local utilities and Best Practices from similar enterprises can also provide ideas for energy projects which are either supported by the local utilities (which means there may be incentives to implement them) or have a strong track record with other similar organizations. Those types of proven projects should be given strong consideration.

13.3   Data to Be Collected

Whatever the tool selected, the information in Table 13.1 will be required for a full analysis (Table 13.1):

The list of information needs per utility and per month may seem overwhelming, but once the initial data is captured, monthly updates are not difficult. On the other hand, if the data is available automatically through Green Button or an automated control or monitoring system, then only the general information, account, and meter details need to be collected manually, and the analysis can be done quickly, potentially providing actionable insights while action is still possible.

The concept of energy use intensity was developed to allow comparisons across different industries, different buildings with varying floor plans, and of course using different utilities which in turn rely on different source fuels. This normalizes the way energy use is compared. The EUI expresses energy use as a function of the building's footprint and is calculated as shown in Fig. 13.1, with an actual calculation in the example. The measure is affected by building type, climate, and other factors, including increased energy effectiveness.

> **Energy Use Intensity Example**
> *A school contains a main floor consisting of 15,000 square feet and a second floor consisting of 10,500 square feet. The school used 1,170,000 kWh of power during the year being analyzed. Kilowatt-hours are multiplied by 3.412 to obtain kBtus; therefore, 1,170,000 × 3.412 = 3,992,040 kBtus. This is divided by the total square footage of 25,500 square feet for an energy use intensity of 3,992,040/22,500 = 156.5 kBtu/sf.*

In Energy Star Portfolio Manager, both site EUI and source EUI are calculated. Site EUI shows the energy use intensity just at the particular facility currently being analyzed, while source EUI also includes losses from the fuel sources and transmission lines that deliver energy to the site. For comparison purposes, source EUI is the better choice.

In Fig. 13.2, a chart drawn from Energy Star Portfolio Manager, supermarkets and groceries have the highest source EUI, while unrefrigerated warehouses have the lowest.

13.4   Financial Analysis

Along with utility data collection and input, the other important analysis is a financial one. Here, management (who know the business) generally have an advantage over consultants (who know the energy industry but often not the business). The basic question to be addressed is how important is energy in the context of the business.

Table 13.1 Data to be collected

General:	
Address	*Water:*
Building square feet	Provider
Building description	Account #(s)
	Meter #(s)
Electricity:	Monthly information (13 months):
Provider	Read date
Account #(s)	Usage
Meter #(s)	Units
Electricity under contract?	Meter size
If so, provider and end date	Fees and other taxes
Rate, variable or fixed	Total for month
Monthly information (13 months):	
Read (or estimate) date	*Sewer:*
Usage/kWh	Provider
Actual kW	Account #(s)
On peak kW	Monthly information (13 months):
Fees and other taxes	Read date
Total for month	Usage
	Units
Natural gas:	Fees and other taxes
Provider	Total for month
Account #(s)	
Meter #(s)	*Stormwater:*
Natural gas under contract?	Provider
If so, provider and end date	Account #(s)
Rate, variable or fixed	Period (days)
Monthly information (13 months):	Total charged
Read date	
Usage	*Other water:*
Units	Total charged
Gas delivery rate	
Fees and other taxes	
Total for month	

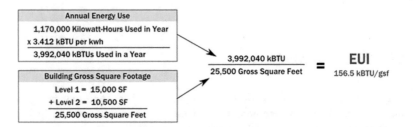

Fig. 13.1 Energy use intensity calculation. (Arch Tool Box (n.d.). https://www.archtoolbox.com/sustainability/energy-use-intensity.html. Accessed 18 July 2017.)

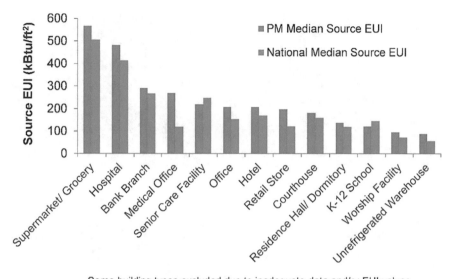

Some building types excluded due to inadequate data and/or EUI values beyond this range

Fig. 13.2 Source energy intensity – median and Portfolio Manager. (Energy Star (n.d.). https://www.energystar.gov/buildings/facility-owners-and-managers/existing-buildings/use-portfolio-manager/understand-metrics/what-energy. Accessed 2 May 2017.)

Illustrative Example, Financial Impact

Consider the income statements for Service Company A, a computer repair facility, and Service Company B, a mobile carpet cleaning company (Fig. 13.3).

Salaries are higher for the computer repair company, and the multiple servers and computers in the building increase electricity costs for the computers and air conditioning, due to the heat the equipment generates. On the other hand, heating costs are low. For the mobile carpet cleaning company, salaries are lower, and electricity is reasonable, but water, natural gas, and fuel are high; the mobile truck is refilled with hot water several times a day at the office, and Company B has a large service territory.

This sort of an analysis would show the managers of Company A that their best cost reduction effort (assuming they are not going to lay people off or decrease their salaries) would focus on reducing electricity use. Company B's managers, on the other hand, would focus on reducing combined heating and water costs, followed by fuel.

(continued)

> *Should either management team be successful in reducing their largest facility-related utility costs by 10%, that reduction would translate to increased profitability. For Company A, there would be a decrease in electrical cost and total expenses of $2200 and therefore a 1% increase in net profit; for Company B the $3900 decrease in natural gas and water costs would increase the net profit by 2%. It would however make little sense for Company A to work on reducing natural gas and water costs nor for Company B to focus on electricity.*

INCOME STATEMENTS			Computer Repair		Mobile Carpet Cleaning	
PERIOD ENDING DECEMBER 31			Service Company A		Service Company B	
INCOME			$	%	$	%
	Services Rendered		$204,000	100%	$ 204,000	100%
EXPENSES						
	Salaries		$100,000	49%	$ 60,000	29%
	Telephone & Internet		$ 6,500	3%	$ 6,500	3%
	Electricity		$ 22,000	11%	$ 6,000	3%
	Natural Gas		$ 2,000	1%	$ 19,000	9%
	Water		$ 1,000	0%	$ 20,000	10%
	Property Rates & Taxes		$ 1,000	0%	$ 500	0%
	Insurance		$ 8,000	4%	$ 8,000	4%
	Advertising / Website Costs		$ 4,000	2%	$ 4,000	2%
	Fuel		$ 1,000	0%	$ 12,000	6%
	Equipment Rental or Depreciation		$ 8,000	4%	$ 8,000	4%
	Bank fees / Interest		$ 500	0%	$ 500	0%
	Professional Fees		$ 1,000	0%	$ 1,000	0%
		Total	$155,000	76%	$ 145,500	71%
NET PROFIT			$ 49,000	24%	$ 58,500	29%

Fig. 13.3 Illustrative income statement

13.5   Business Analysis

The first part of the SEE Framework focused on the value and purpose in a strategic sense. The external assessment must also make that connection, linking to the organization's mission and structure. Energy projects should be implemented in ways that improve the organization's strategic positioning and model.

> - *Returning to Companies A and B referenced in Sect. 13.4, Service Company B is a "mobile carpet cleaning" company, and it must remain "mobile" or change its business model. Reducing fuel costs by selling the vehicle providing that mobile service would indeed cut fuel costs – but it would not support the strategy or improve the company's ability to provide services.*
> - *An organization which has as its core mission the analysis and reselling of eco-friendly building materials to customers on-site would improve the consonance between its services and image if the building were constructed of eco-friendly materials. If that were so, and if the building design showcased some ways the materials could be used, the building could also be used as a sales tool.*

References

Arch Tool Box (n.d.). *Energy use intensity.* Retrieved from Archtoolbox.com, Architect's Technical Reference: https://www.archtoolbox.com/sustainability/energy-use-intensity.html

Baker, N. E. (2017). *Institutional change federal energy management.* Retrieved from Energy.gov/eere/femp: https://energy.gov/eere/femp/institutional-change-federal-energy-management

Energy Star (n.d.). *Use portfolio manager/understand metrics/what energy.* Retrieved from www.energystar.gov/buildings/facility owners and managers: https://www.energystar.gov/buildings/facility-owners-and-managers/existing-buildings/use-portfolio-manager/understand-metrics/what-energy

Hansen, S. (n.d.). *Making the business case for energy efficiency.* Retrieved from docplayer.net: http://docplayer.net/6344792-Making-the-business-case-for-energy-efficiency-shirley-j-hansen-ph-d.html

Studney, C. (2012, October 4). *How to measure the ROI of LEED.* Retrieved from www.greenbiz.com/blog: https://www.greenbiz.com/blog/2012/10/04/how-measure-roi-leed

Chapter 14
Goals

14.1 Introductory Comments

It is important to set goals before beginning the process of internal assessment for an energy effectiveness *project* (a one-time initiative which often requires investment) or *program* (an extended effort planned to be continued over a period of time. Goals set parameters around the internal assessment, so that when it is conducted, it will be focused upon the strategic issues important to the company and targeted to collect the relevant information most efficiently. During the initial external assessment, tentative targets may have been set in relation to competitors, but during this stage of the process, those tentative targets should be reexamined and adjusted if necessary.

Energy efficiency programs, alternative energy projects, and all similar activities should improve the company and its shareholders, staff, and communities. They should be connected to the strategy and should help to improve the company's strategic positioning and the company as a whole. The financial goals for the enterprise should be those used to choose between the energy projects or programs and other alternatives, because energy projects are no different from other projects in their potential claim on scarce resources. All projects within an organization should meet the same goals for revenue enhancement, cost displacement, quality improvement, and other parameters for improvements.

14.2 Goals as Filters for Decision-Making

Goals are set in order to determine what needs to be achieved but also to provide a way to choose between alternative scenarios, filtering out those which do not meet organizational goals. For energy projects, they are often the first filter; technical, financial, and organizational filters follow in subsequent stages of the SEE Framework.

© Springer International Publishing AG, part of Springer Nature 2018
S. McCardell, *Energy Effectiveness*, https://doi.org/10.1007/978-3-319-90255-5_14

Goals as a Filter: Example

A small town has a municipal utility which provides potable water to its residents. Over the last several years, the water company's infrastructure has shown increasing signs of age, but the town has been growing and has managed to put aside some funds to cover infrastructure improvements; in addition, they have unused bonding capacity, so they feel comfortable that they can begin a phased upgrade program. They are considering three different projects: pump replacement/repair, a new training center, and smart meter installation.

Management feels that the upgrades will enable them to reduce the ongoing risk of catastrophic failure and meet their commitments to serve their customers, and therefore, they have to prioritize and schedule them. They complete the following qualitative analysis (Table 14.1).

The installation costs for each option are assumed to be equal at $100,000, and for the purpose of this exercise, the anticipated savings are assumed to be equivalent as well. The correct option would therefore be the one which is most closely aligned with the municipal utility's purpose, goals, and strategies.

Table 14.1 Qualitative Project Comparison

	Pump replacement/repair	New training center	Meter installation
Benefits	Avoid disruption	Replace leaking roof	Reduce non-revenue water
	More efficient	Host regional classes	Reduce meter reader winter travel
	Smaller space	Benefit to community	Identify leaks
Negatives	Timing not good	Old building well-loved	Customer disruption
	Maintenance unfamiliar	Reduces view from Senior Ctr	Customer concerns
	Pipes may fail		Training
Savings	10% on electricity	Catering fees	Prevent leaks
	50% on maintenance	Room space rental	Increase revenue
	5% on non-revenue water	Transportation cost	Reduce "drive around time"
Costs	Purchase & installation	Demolition & design	Meter purchase
	Training	Construction	Meter installation
	Pipe replacement	Internal furnishings	Software
Investment	$100,000	$100,000	$100,000

14.3   Goals Expressed in Financial Terms

There are substantial, deep, and long-term benefits of sustainable and energy-effective equipment purchases, but as a rule, neither benefits nor costs are adequately accounted for, and it is with difficulty that they can be connected to goals. The simple example below highlights this point.

> **Goals Made Real: Example**
> *For example, let's consider a small preschool which budgeted $3000 to construct a playground which they believed would keep their small charges happier and healthier during the day, enabling them to pay more attention when they were in the classroom. Summer came, and although their antici-pated enrollment for the following year was somewhat lower than expected, they constructed the playground anyway. As it happened, it was a great play-ground, and several neighbor families who had not previously considered the school enrolled their children, seeing that the school really valued their students.*
>
> *A similar preschool in the adjoining town was also contacted by the playground designer, and although they were doing quite well in terms of enrollment and had the resources available, they chose not to spend $3000 on the playground because they felt they needed to increase their advertising/ marketing program while they had the funds to do so.*

In this sort of a scenario, one could argue that either or both schools made the right decision – but from the information presented, it appears that the first pre-school chose an option more closely aligned with its purpose, a decision that was recognized by potential parent/clients.

14.4   Goals Which Can Be Managed and Measured

Once the enterprise's important goals are identified, the next step is to determine how progress toward those goals will be achieved, measured, and managed as an ongoing part of operations. The implementation process will be discussed later, but the management and measurement methodologies should be considered at the same time as the goals; the same is true of planning how to include energy project results in standard business metrics. This is a critical step in the transi-tion from consideration of energy challenges and opportunities to actions and impacts. Only what is measured, as has already been noted, can be managed and improved.

Depending on the project or program, different types of measurement or mea-surement and verification (if an outside entity requires verified results) may be appropriate.

- A lighting retrofit in a large industrial facility with a dedicated electrical box for the new lighting can be measured in various ways:
  - Calculations based on the decreased wattage, adjusted for assumed usage
  - Installation of a sensor to measure and record electricity flow inside the lighting connection box

- Connecting that same sensor to a monitoring system or even perhaps a phone app, which could provide graphical data and alerts as well as other details such as building occupancy
- Adding a lighting control system for even greater functionality beyond strict measurement, potentially including daylight harvesting capability for sunny days, occupancy sensors, and fault sensing

Any one of these measurement methodologies might work; the right choice depends on the situation and the organization.

Most managers will comment that the resource which is in shortest supply is time. This is important to remember when choosing measurement methodologies, because as previously discussed any project results that are not measured are not likely to be managed, either. If a junior staffer is assigned the task of calculating the lighting savings weekly or monthly and then overwhelmed with more time-critical projects, the information will not be generated, and the project will descend into invisibility again.

14.5 Goals Connected to Key Performance Indicators

Goals should also be connected to management's key performance indicators, or KPIs, to bring the energy project results into the fabric of the enterprise. The concept of KPIs is extremely useful, and even before the term was coined, most managers made use of the concept. Investors looked at ratios of debt to equity as a quick guide to the financial health of a company, manufacturing managers considered throughput or percentages of defects, and partners in law firms kept track of ratios of billable hours to total hours. KPIs combine important metrics in ways that give managers a quick feel for the performance of an organization over time.

For electricity, natural gas, and water, there are several steps to take in developing an appropriate KPI.

1. Consider whether any KPIs currently used by the enterprise could be adapted to incorporate energy costs as well.
2. Track the price of the commodity separately, but use usage and demand fees or peak charges in the KPI because those are the elements most subject to management's control.
3. Estimate where the most electricity, natural gas, and water are used (on the assumption that all utilities are under investigation), through each of the Four Fields (buildings, equipment and processes, environment, and people), for example:

   (a) Is water part of the manufacturing process? If so, then water and sewer use should be correlated with manufacturing output.
   (b) Is heating a significant issue? If so, then natural gas costs should be adjusted for outdoor temperature and, if thermostat settings are variable, building occupancy.

(c) Are personal computers a significant power draw? If so, then relating that usage to the average number of people and occupancy hours would be appropriate.

Other potential relationships might be the types of products produced on a manufacturing line, the square feet of occupied space, the hours a particularly large piece of equipment or energy-intensive process is running, the gallons of product produced, the number of people per shift, and others. If management goes through this process thoughtfully, many of the relationships between different parts of the organization will become clear, and control parameters can be developed.

Only in rare cases are utility usage patterns consistent from month to month. There is seasonal variation of course, along with changes in product mix, services provided, orders, staff, and other elements. Whether utility usage and costs are being tracked on paper by looking at utility bills and calculations, or by controls and monitors, it's important to construct averages. Utilities can be tracked using a 12-month rolling average, comparing EUI or energy use intensity, total energy usage over a week at different points during the year, or average demand/peak use charges per month.

14.6   Goals as Coordinated and Managed by the Champion

As most managers can attest, very little is done without what is called in the industry a champion. The champion may be an individual or a team charged with driving interest in the energy project, soliciting input from others, advocating for the project, and in other ways supporting the energy initiative. The champion also ensures that the project's connection to the organization's goals remains strong and that the projects are put through the same rigorous analysis required for any project proposed to management. This champion need not be the CEO/COO but should be a credible professional, have the ear of top management, and be able to call in additional resources when necessary.

References

American Council for an Energy Efficient Economy (ACEEE) (n.d.). *Programs*. Retrieved from ACEEE.org: http://aceee.org/portal/programs
Baker, N. E. (2017). *Institutional change federal energy management*. Retrieved from Energy. gov/eere/femp: https://energy.gov/eere/femp/institutional-change-federal-energy-management
Bertoldi, P. E. (2016, June). *Energy efficiency, volume 09, issue 03*. Dordrecht: Springer Publishing.
Carayannis, E. E. (2014). *Business model innovation as lever of organizational sustainability*. New York: Springer Science+Business Media. https://doi.org/10.1007/s10961-013-9330-y.
Diana, F. (2011, March 9). *The evolving role of business analytics*. Retrieved from Frank Diana's Blog – Our Emerging Future: https://frankdiana.net/2011/03/19/the-evovling-role-of-business-analytics/

Economist Intelligence Unit. (2012). *Energy efficiency and energy savings; a view from the building sector*. London: The Economist Intelligence Unit LTD.

Eggink, J. (2007). *Managing energy costs: A behavioral and non-technical approach*. Lilburn, GA: Fairmont Press.

Energy Information Administration (n.d.). *Today in energy*. Retrieved from EIA.Gov: https://www.eia.gov/todayinenergy/detail.php?id=18071

Energy Star (n.d.). *Use portfolio manager/understand metrics/what energy*. Retrieved from www.energystar.gov/buildings/facility owners and managers: https://www.energystar.gov/buildings/facility-owners-and-managers/existing-buildings/use-portfolio-manager/understand-metrics/what-energy

Hansen, S. (n.d.). *Making the business case for energy efficiency*. Retrieved from docplayer.net: http://docplayer.net/6344792-Making-the-business-case-for-energy-efficiency-shirley-j-hansen-ph-d.html

Leonardo Academy (n.d.). *Free courses and programs on sustainable energy*. Retrieved from Leonardo Academy: www.leonardo-academy.org

Mourik, R. E. (2015). *What job is Energy Efficiency hired to do? A look at the propositions and business models selling value instead of energy or efficiency*. Retrieved from IEADSM Leonardo Energy: https://www.youtube.com/watch?v=GGLYp_fHrMs

New Buildings Institute (2017, July 15). *Deep energy retrofits*. Retrieved from New Buildings Institute: http://newbuildings.org/hubs/deep-energy-retrofits

Powerhouse Dynamics (2017). *Solutions – screen captures sent via personal communications*. Retrieved from www.poerhousedynamics.com: https://powerhousedynamics.com/solutions/sitesage/

Public Technology Inc./US Green Building Council. (1996). *Sustainable building technical manual*. Washington, DC: Public Technology.

Russell, C. C. (2010). *Managing energy from the top down; connecting industrial energy efficiency to business performance*. Lilburn, GA: The Fairmont Press Inc./CRC Press.

Shields, C. (2010). *Renewable energy facts and fantasies*. New York: Clean Energy Press.

SustainAbility (2014). 20 Business model innovations for sustainability.. SustainAbility.

The Shift Project (2012). *Top 20-capacity chart*. Retrieved from http://www.tsp.org The Shift Project: http://www.tsp-data-portal.org/TOP-20-Capacity#tspQvChart. Accessed 17 Feb 2017.

US Department of Energy: Energy Efficiency and Renewable Energy (2011). *A guide to energy audits PNNL-20956*. Pacific Northwest National Laboratory. Retrieved from http://www.pnnl.gov/main/publications/external/technical_reports/pnnl-20956.pdf

US Energy Information Agency (n.d.). *faqs*. Retrieved from eia.gov: https://www.eia.gov/tools/faqs/faq.php?id=427&t=3

US Environmental Protection Agency (2017, January 24). *Energy and the environment/electricity – customers*. Retrieved from www.epa.gov: https://www.epa.gov/energy/electricity-customers

Woodroof, E. A. (2009). *Green facilities handbook: Simple and profitable strategies for managers*. Lilburn, GA: The Fairmont Press.

World Business Council for Sustainable Development (n.d.). *Making tomorrow's buildings more energy efficient*. Retrieved from www.wbcsd.org: http://www.wbcsd.org/Overview/Resources

World Economic Forum (2017). *Reports*. Retrieved from weforum.org: http://reports.weforum.org/global-energy-architecture-performance-index-report-2016/energy-access-and-security/. Accessed 6 July 2017.

Chapter 15
Internal Assessment

15.1 General Comments

This brief summary chapter focuses on assessing the situation inside the enterprise, as structured by the external assessment and the organization's goals. Technical specifics fall within the Fields of building, equipment/process, and environment and they are generally included in energy audits. In addition, the financial and human factor tools described in Chap. 9 expand the analysis to one which includes the entire enterprise. Within the structured framework, the emphasis is on the organizational assessment rather than on the technical details which would be more appropriately considered by outside engineering consultants.

15.2 Energy Audits and Monitoring/Control Systems: A Note

Energy audits, mentioned in Chap. 6, will not be discussed further although they can provide an excellent snapshot of the facility at a particular point in time as well as snapshots of longitudinal data over extended periods. There is an extensive body of knowledge related to energy audits, including a few guides referenced in Chap. 21. Sensors, monitors, software, and control systems provide a better understanding of ongoing operations, as well as the capability to manage energy use and cost in real time. They should be installed and used whenever feasible. Although the SEE Framework does not depend on such systems, energy management is much easier, relevant, efficient, and effective with them installed. They will therefore be assumed for the remainder of this book.

© Springer International Publishing AG, part of Springer Nature 2018
S. McCardell, *Energy Effectiveness*, https://doi.org/10.1007/978-3-319-90255-5_15

15.3   Assessment Methodology

As the third segment in the strategic framework process, the internal assessment relies on the information developed and collected in the two previous segments. The focus of the assessment is on the portion of the utilities that contribute to productive use, which is the work those utilities are intended to accomplish.

The types of questions to which answers should be found at this stage relate to productive use, formal and informal internal organization structures, and staff concerns/involvement. Detailed information on the facility and utility accounts will already have been collected.

15.4   Productive Use in the Four Fields

In each of the Four Fields, the best way to conceptualize how utilities are used and how divisions work together in a system is to focus on flows. How do electricity, natural gas, and water flow? How do investments flow, as well as savings? How and where are decisions made? And then how are all of those connected to the rest of the organization?

As a reminder, the Four Fields consist of:

- Building
- Environment
- Equipment, processes, and procedures

    - Manufacturing processes
    - Business procedures

- People

15.5   First Pass Qualitative Questions/Four Fields

The assessment for the first three fields is very similar, beginning with general organization data collection which will be primarily qualitative at this point.

- Who owns the building? What are the terms if the building is leased or rented?
- What senior management/CFO concerns exist?
- What is the general interest in energy effectiveness, and why?
- What is the general interest in sustainability, and why?
- What are the most common decision-making methodologies (formal/informal/by type of project)?
- What is the desired ROI for projects?
- Are there any energy- and water-related concerns?

- Are there any operational opportunities and/or concerns?
- What energy conservation measures have been taken?
- What measures are planned?
- In what ways does this field contribute most importantly to the organization's value or purpose?
- What productive use is expected from electricity, natural gas, water, or any other utilities?
- Where are electricity, natural gas, and water used?
- What measures have been taken to be more energy efficient or conserve energy/ water?
- What does energy use vary with?

  – What patterns can be discerned that relate to those items?

- How are utilities allocated?
- What is the management and accountability structure?
- How do communications flow?
- What should be measured, and who would monitor results?

The people field, however, is quite different. Each individual in an organization is important not only because of their individual areas of influence but also because each has the capability to impact energy and water usage dramatically. Questionnaires are not the best way to address issues in the people field because both formal and informal structures are important. The tools and rules from the five-step framework provide a good structure to analyze the people field and will also help address business procedures.

Additional checklists can be found in Chap. 20, although these are likely to require adaptation for application in a particular enterprise.

References

American Council for an Energy Efficient Economy (ACEEE) (n.d.). *Programs*. Retrieved from ACEEE.org: http://aceee.org/portal/programs

Baker, N. E. (2017). *Institutional change federal energy management*. Retrieved from Energy. gov/eere/femp: https://energy.gov/eere/femp/institutional-change-federal-energy-management

Bertoldi, P. E. (2016, June). *Energy efficiency, volume 09, issue 03*. Dordrecht: Springer Publishing.

Carayannis, E. E. (2014). *Business model innovation as lever of organizational sustainability*. New York: Springer Science+Business Media. https://doi.org/10.1007/s10961-013-9330-y.

Diana, F. (2011, March 9). *The evolving role of business analytics*. Retrieved from Frank Diana's Blog – Our Emerging Future: https://frankdiana.net/2011/03/19/the-evovling-role-of-business-analytics/

Economist Intelligence Unit. (2012). *Energy efficiency and energy savings; a view from the building sector*. London: The Economist Intelligence Unit LTD.

Eggink, J. (2007). *Managing energy costs: A behavioral and non-technical approach*. Lilburn, GA: Fairmont Press.

Energy Information Administration (n.d.). *Today in energy*. Retrieved from EIA.Gov: https://www.eia.gov/todayinenergy/detail.php?id=18071

Energy Star (n.d.). *Use portfolio manager/understand metrics/what energy.* Retrieved from www. energystar.gov/buildings/facility owners and managers: https://www.energystar.gov/buildings/ facility-owners-and-managers/existing-buildings/use-portfolio-manager/understand-metrics/ what-energy

Hansen, S. (n.d.). *Making the business case for energy efficiency.* Retrieved from docplayer.net: http://docplayer.net/6344792-Making-the-business-case-for-energy-efficiency-shirley-j-hansen-ph-d.html

Leonardo Academy (n.d.). *Free courses and programs on sustainable energy.* Retrieved from Leonardo Academy: www.leonardo-academy.org

Mourik, R. E. (2015). *What job is Energy Efficiency hired to do? A look at the propositions and business models selling value instead of energy or efficiency.* Retrieved from IEADSM Leonardo Energy: https://www.youtube.com/watch?v=GGLYp_fHrMs

New Buildings Institute (2017, July 15). *Deep energy retrofits.* Retrieved from New Buildings Institute: http://newbuildings.org/hubs/deep-energy-retrofits

Powerhouse Dynamics (2017). *Solutions – screen captures sent via personal communications.* Retrieved from www.poerhousedynamics.com: https://powerhousedynamics.com/solutions/ sitesage/

Public Technology Inc./US Green Building Council. (1996). *Sustainable building technical manual.* Washington, DC: Public Technology.

Russell, C. C. (2010). *Managing energy from the top down; connecting industrial energy efficiency to business performance.* Lilburn, GA: The Fairmont Press Inc./CRC Press.

Shields, C. (2010). *Renewable energy facts and fantasies.* New York: Clean Energy Press.

SustainAbility (2014). 20 Business model innovations for sustainability.. SustainAbility.

The Shift Project (2012). *Top 20-capacity chart.* Retrieved from http://www.tsp.org The Shift Project: http://www.tsp-data-portal.org/TOP-20-Capacity#tspQvChart. Accessed 17 Feb 2017.

US Department of Energy: Energy Efficiency and Renewable Energy (2011). *A guide to energy audits PNNL-20956.* Pacific Northwest National Laboratory. Retrieved from http://www.pnnl. gov/main/publications/external/technical_reports/pnnl-20956.pdf

US Energy Information Agency (n.d.). *faqs.* Retrieved from eia.gov: https://www.eia.gov/tools/ faqs/faq.php?id=427&t=3

US Environmental Protection Agency (2017, January 24). *Energy and the environment/electricity – customers.* Retrieved from www.epa.gov: https://www.epa.gov/energy/electricity-customers

Woodroof, E. A. (2009). *Green facilities handbook: Simple and profitable strategies for managers.* Lilburn, GA: The Fairmont Press.

World Business Council for Sustainable Development (n.d.). *Making tomorrow's buildings more energy efficient.* Retrieved from www.wbcsd.org: http://www.wbcsd.org/Overview/Resources

World Economic Forum (2017). *Reports.* Retrieved from weforum.org: http://reports.weforum. org/global-energy-architecture-performance-index-report-2016/energy-access-and-security/. Accessed 6 July 2017.

Part V
Introducing and Using the Strategic Energy Effectiveness Framework (SEE): Planning, Implementation, Adaptation

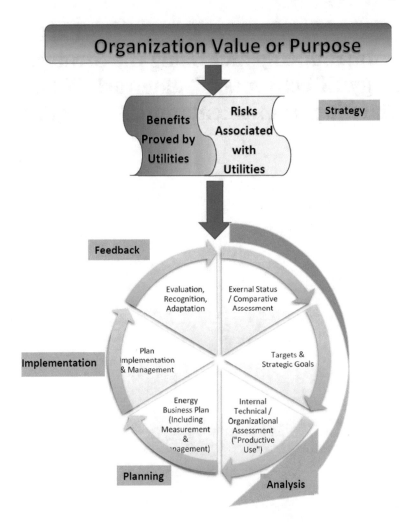

Part C of the Strategic Energy Management Framework includes the three phases of Planning, Implementation, and Feedback or Adaptation as shown in Fig. 1. The second half of the Framework is oriented towards action.

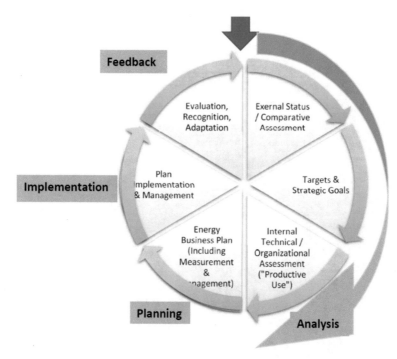

Fig. 1 Planning, Implementation, and Adaptation in the SEE Framework. (Copyright 2016–2018 Current-C Energy Systems, Inc. Used with Permission)

Chapter 16
Energy Business Plan

16.1 What Results Could Be Achieved or Expected?

Planning and implementation for all projects and programs is a challenge, and technical complications appear to make energy projects yet more difficult. Financial concerns and organizational imperatives and constraints are however key for energy projects as they are for all other initiatives; technical considerations are secondary and that expertise can be provided by other experts. To highlight this point, we use the term Energy Business Plan.

Potential savings and appropriate energy projects depend to a large extent on the industry and the particular circumstances of a specific enterprise. For industries with highly energy- and water-intensive usage, the solution is likely to be a technical one, using more efficient technologies or equipment to reduce costs and usage. Where energy and water use are variable over time, by product, or in other unpredictable ways, behavior change and programs focused on people are the more logical choice.

> **Example: An energy effective enterprise:** *At one end of the scale might be a warehouse operation which just moved into a new passive solar building in a mid-sized town where many employees commute by bus or by bicycle. Deliveries to and from the facility rely on an adjoining railway freight line. The building is 20% office space and 80% un-refrigerated warehouse. Power is provided by a grid-tied solar system that provides for 85% of the power requirements, which are low in aggregate because skylights and daylighting harvesting strategies reduce lighting costs, and a ground-source heat pump system serves to both heat and cool the building. Fork lifts are electric, and although they are charged during the night when the solar panels are not*

(continued)

© Springer International Publishing AG, part of Springer Nature 2018
S. McCardell, *Energy Effectiveness*, https://doi.org/10.1007/978-3-319-90255-5_16

working, the utility rates at night are about half the rates during the day. In addition, for about 5 months a year, the solar system produces more power than the warehouse needs, so that excess is fed back to the utility as a credit to the warehouse account. Management calculates that although they do purchase 15% of their power from the utility, they pay approximately 20% less than they would if they purchased the same number of kWh on a normal operating schedule. Warehouse staff also enjoy the use of a garden landscaped with local plantings and trees, all of which are watered by a system that recycles water from the facility. Staff are enthusiastic about these cost reduction strategies which they also see as amenities and are committed to maintaining them. In fact, management has noticed that several employees check the real-time energy monitoring graphs on the energy and water management system and report anomalies to maintenance personnel.

This particular enterprise is highly energy effective. Alternative energy systems, energy-efficient equipment and strategies, and an organization-wide commitment to energy and water issues suggest that the warehouse could only reduce costs at the margin and on the order of a few percentage points, most likely through actions suggested or implemented by staff who compete to make useful suggestions.

Example: An enterprise with opportunity to become more energy effective: *At the other end of the scale might be a cooperative furniture-making operation located in a mountainous area. The members of the cooperative (who are also the managers) moved the company to its current location because it is near where they live, but recently when the electric utility instituted a large rate increase to cover the cost of replacing aging transmission and distribution lines, they wondered if that move had been a mistake – the rate increase would reduce their profit margin by almost 2%, since the facility is all electric powered. The company was in fact already frustrated because the utility service was unreliable (which was of course the reason for the planned line replacement). Feeling as though this was the last straw, the cooperative decided to look for other alternatives. With grant and loan funding from suppliers and government agencies as well as retained earnings and their own time and energy, they installed a system with the following components:*

- *Energy-efficient LED lighting*
- *Various load management devices on all fans, motors, and other equipment, including variable speed and/or variable frequency drives*
- *Hybrid electricity generation system which drew from the nearby year-round stream and two wind turbines situated in an adjoining meadow that had been recently cleared, along with a battery backup system*

(continued)

> *Once the project was complete, the furniture-making cooperative disconnected from the utility, relying entirely on its own generation capacity and internal processes and procedures intended to achieve maximum energy effectiveness. Using these strategies, the organization reduced their outside energy use by 100%.*

The furniture-making operation took advantage of every opportunity they could find, in creative ways, to reduce their utility costs and solve their power reliability problem.

16.2   Strategies to Consider as Part of the Energy Business Plan

If those two examples might be considered ends of the scale, most organizations would fall between them depending on the constraints inherent in the original design elements found in the Four Fields.

The strategies appropriate for each should flow from the analyses completed in the first sections of the Strategic Framework which set out the external assessment, the organizational goals, and the internal assessment. Types of strategies that might be considered, and the estimated savings potential, are described below.

- Building

  - Buildings have inherent energy efficiency properties, some of which can be more easily changed than others. Metal buildings, for example, are generally uninsulated but can have insulation added either inside or outside the metal shell. It is more difficult to do the same for historic structure of wood, stone, adobe, plaster, or other materials – particularly if there are controlling requirements from historic associations.
  - One of the tenets of green building design is that the most important determinant of the energy efficiency of a building is the way it sits on the site. Window placement, relationship to neighboring buildings, and other similar components are very difficult to change.

- Equipment, Processes, and Procedures

  - Equipment

    - Operations which use large pieces of equipment are similarly constrained; a wastewater plant relies on numerous pumps to move, mix, and aerate the wastewater. Those pumps cannot be taken out, but they can be replaced with more efficient and better ones.

– Manufacturing processes

  • If an enterprise is involved in manufacturing, then the energy requirements are likely to be high, and there is little chance that the manufacturing line can be removed; it can generally be improved, however.

– Business procedures

  • Business procedures are less visible than manufacturing processes, but sometimes they are no less solid or resistant to change. An inventory control system, for example, might require individually molded plastic tags for each item in inventory, physical inspection, and manual input. While there are dozens of ways to continue to control inventory, improvements in these systems might also increase energy effectiveness; designing such adaptations represents a difficult challenge.

• Environment

  – The ability to source water, power, and heat has freed people from the necessity to take the environment into consideration in designing and locating buildings.
  – The environment has a strong influence on the energy that is required to make occupants comfortable, however, and that environmental impact cannot be ignored; it can only be minimized. Imagine staying in a hotel in Dubai in the United Arab Emirates, or one in Oslo in Norway, without energy for cooling, heating, or water.

• People

  – Theoretically, people represent the part of the organization which can change most easily and most completely, with the least amount of effort. Neither cultural change nor organizational change is easy, however, and increasing energy effectiveness requires both, in a sustained manner.

16.3   Categories of Opportunities

Potential savings then relate to the enterprise's fundamental business model, the place the facility is located, the building in which it is housed, and people's openness to change. There are two basic categories of opportunities:

Step-change or incremental improvements can come from approaches such as:

– Adjusting processes and procedures
– Adjusting inputs
– Reducing and recapturing waste
– Revising specifications for additional equipment purchases

Substantially game-changing improvements can come from approaches such as:

- Installing real-time monitoring capability including access for all employees
- Leadership
- Rethinking ways in which the enterprise can accomplish its business purpose
- Cascading use of waste and process heat, reusing it as often as possible
- Examining each energy-related investment decision using standard financial analytics and comparing those proposed projects to other potential uses of funds
- Deep energy retrofits where multiple technologies and strategies are implemented simultaneously
- Helping employees learn that everything they do has an impact, see the benefit of their actions and be recognized for them, and providing incentives for further improvement

Within those two general categories, there are essentially five patterns into which energy projects tend to fall:

- *Engineering-led projects* focus on structural changes to equipment and manufacturing, HVAC, process, or controls systems.

  - Projects may require investment or partnerships with energy services companies (ESCOs) which design and install, maintain, and handle operations and maintenance for a negotiated percentage of savings.
  - Projects usually do not include analysis of or savings from changes in how energy and water are used and controlled daily.
  - With monitors or controls systems, results can be captured easily.
  - Savings can be up to 10–20% per year for each utility cost category targeted.

- *Low-cost/no-cost projects*, generally used when companies are new to the idea of energy effectiveness and want to try it first or have few investment resources.

  - These types of projects focus on quick wins such as running existing equipment and processes more efficiently on a day to day basis, staff training to identify and capture savings in operations or maintenance, and behavior change programs.
  - The concept is to identify a low-risk but well-proven approach with projected savings that also contributes to reduced maintenance and/or safety. Then monitor the result, and incrementally expand the project scope as that becomes possible.

    - Examples include changing HVAC set points and programs so that heating and cooling systems do not operate simultaneously and against one another, setting up a system for rapid reporting and repair of maintenance issues, or asking the last person to leave at night to turn off all equipment and lights still running and leave a report at the front desk.

- Savings increase in aggregate over the period of ongoing incremental projects, with each one contributing around 1% savings and maximum savings of around 5–10%. Continued emphasis on the programs is required for the savings to persist.

- *Process, procedure, and systems improvements* may not require large investment, but they can take a great deal of staff and contractor time in the design phase.

  - These types of systems improvements often have long-term benefit to the enterprise beyond energy or water savings and may improve its competitive position.
  - Examples include implementing more effective maintenance procedures including (for lighting) group re-lamping standards, putting in place reporting systems that relate energy use, energy bills, and energy payment, installing moisture meters at selected points on an irrigation system, and installing an energy monitoring system.
  - As may by this time be clear, these types of systems, process, and procedure improvements are key to increasing energy effectiveness.
  - Energy savings (excluding significant potential savings in other areas) for these types of improvements can be 10–20%.

- *Structural opportunities* can be addressed by large investment projects which address, to the extent appropriate, the Four Fields constraints.

  - These projects require specific technologies, building systems, or equipment and may sometimes be installed more cost-effectively in combination.
  - Examples include a full LED lighting retrofit throughout a large facility, redesigning a building layout or production line, installing an alternative energy system such as solar power generation or a combined heat power system, and upgrading manufacturing equipment.
  - Savings can be 20–40%, and the return on investment can be very good for projects which are well designed.

- *Core redesign projects* take a deep look at the business to understand whether new business models, new technologies, and other similar radical changes could improve energy effectiveness.

  - These are generally strategic efforts driven by concerns other than energy, and they have the potential to bring radical improvements in energy effectiveness.
  - Where increased market and strategic advantages are sought, increased resource efficiency should also be considered since that may enable the enterprise to more successfully achieve the desired result.
  - Examples include changing the business model in some major way and redesigning products or services.
  - Savings can be 20–50% when a significant investment is made and spread across the entire organization. Where circumstances permit consideration of a

more fundamental look at the business model, radical energy improvements with extraordinary returns may be possible. (Usually these types of projects are undertaken to achieve market/strategic advantages as well as improved resource productivity.)

The potential reductions in energy intensity are of course only estimates and furthermore should not be summed. There are several reasons for this.

- Each organization and situation is unique, and not all approaches can be used for any particular enterprise.
- Implementation of one project may preclude implementation of another; for example, replacing all fluorescent with LED lights and a control system (a highly efficient option) would remove the opportunity to replace the original fluorescents with somewhat more efficient fluorescent models and individual room occupancy sensors.
- Savings calculations, even when very rough as in the percentages above, use the current cost as a base. If multiple projects are installed, each of them will reduce the base load for the next, and therefore the aggregated savings percent will be less than that for the projects individually.

16.4   Developing the Energy Business Plan

In the SEE Framework, the project plan is called an Energy Business Plan because the same types of considerations should be part of this kind of comprehensive project as are part of a business plan. Completion of previous segments of the Framework wil have generated the information required to produce such a plan.

For enterprise-driven projects, the best approach is to start at the base of the pyramid of opportunities shown in Fig. 16.1 and move up the pyramid as appropriate, because not all levels will be relevant to every facility. Note that the lower levels of the pyramid require less investment than the higher levels but that the return on investment can be high at each stage.

16.5   Energy Program and Project Success

A great deal of research has been conducted and papers written about what makes an energy project design successful. At this point in the book, none of the elements of success should be surprising, but they are important to remember as planning proceeds.

- Steering committee.

Fig. 16.1 Pyramid of opportunities. (Copyright 2015–2018 Current-C Energy Systems, Inc., Used with Permission.)

- Training, metrics and reporting, raising energy IQ, suggestion box input, continually improve and implement more training
- The organization and management should make a serious commitment to become more energy effective.
- Commitment and involvement should start at the executive level and continue throughout the organization.
- Systems of accountability and appropriate rewards should be embedded.
- If possible, start with easy efforts.
- Goals should relate to vision and purpose.
- Energy programs should be analyzed using the organization's standard project evaluation criteria.
- Goals that can and will be measured but are not exclusively financial or energy-related should be set.

- Performance should be measured over time.
- Communicate and take action; empower staff – give information, engage in initiatives, and celebrate achievements.

References

American Council for an Energy Efficient Economy (ACEEE) (n.d.). *Programs*. Retrieved from ACEEE.org: http://aceee.org/portal/programs

Baker, N. E. (2017). *Institutional change federal energy management*. Retrieved from Energy. gov/eere/femp: https://energy.gov/eere/femp/institutional-change-federal-energy-management

Bertoldi, P. E. (2016, June). *Energy efficiency, volume 09, issue 03*. Dordrecht: Springer Publishing.

Carayannis, E. E. (2014). *Business model innovation as lever of organizational sustainability*. New York: Springer Science+Business Media. https://doi.org/10.1007/s10961-013-9330-y.

Diana, F. (2011, March 9). *The evolving role of business analytics*. Retrieved from Frank Diana's Blog – Our Emerging Future: https://frankdiana.net/2011/03/19/the-evovling-role-of-business-analytics/

Economist Intelligence Unit. (2012). *Energy efficiency and energy savings; a view from the building sector*. London: The Economist Intelligence Unit LTD.

Eggink, J. (2007). *Managing energy costs: A behavioral and non-technical approach*. Lilburn, GA: Fairmont Press.

Energy Information Administration (n.d.). *Today in energy*. Retrieved from EIA.Gov: https://www.eia.gov/todayinenergy/detail.php?id=18071

Energy Star (n.d.). *Use portfolio manager/understand metrics/what energy*. Retrieved from www.energystar.gov/buildings/facility owners and managers: https://www.energystar.gov/buildings/facility-owners-and-managers/existing-buildings/use-portfolio-manager/understand-metrics/what-energy

Hansen, S. (n.d.). *Making the business case for energy efficiency*. Retrieved from docplayer.net: http://docplayer.net/6344792-Making-the-business-case-for-energy-efficiency-shirley-j-hansen-ph-d.html

Leonardo Academy (n.d.). *Free courses and programs on sustainable energy*. Retrieved from Leonardo Academy: www.leonardo-academy.org

Mourik, R. E. (2015). *What job is Energy Efficiency hired to do? A look at the propositions and business models selling value instead of energy or efficiency*. Retrieved from IEADSM Leonardo Energy: https://www.youtube.com/watch?v=GGLYp_fHrMs

New Buildings Institute (2017, July 15). *Deep energy retrofits*. Retrieved from New Buildings Institute: http://newbuildings.org/hubs/deep-energy-retrofits

Powerhouse Dynamics (2017). *Solutions – screen captures sent via personal communications*. Retrieved from www.poerhousedynamics.com: https://powerhousedynamics.com/solutions/sitesage/

Public Technology Inc./US Green Building Council. (1996). *Sustainable building technical manual*. Washington, DC: Public Technology.

Russell, C. C. (2010). *Managing energy from the top down; connecting industrial energy efficiency to business performance*. Lilburn, GA: The Fairmont Press Inc./CRC Press.

Shields, C. (2010). *Renewable energy facts and fantasies*. New York: Clean Energy Press.

SustainAbility (2014). 20 Business model innovations for sustainability.. SustainAbility.

The Shift Project (2012). *Top 20-capacity chart*. Retrieved from http://www.tsp.org The Shift Project: http://www.tsp-data-portal.org/TOP-20-Capacity#tspQvChart. Accessed 17 Feb 2017.

US Department of Energy: Energy Efficiency and Renewable Energy (2011). *A guide to energy audits PNNL-20956.* Pacific Northwest National Laboratory. Retrieved from http://www.pnnl. gov/main/publications/external/technical_reports/pnnl-20956.pdf

US Energy Information Agency (n.d.). *faqs.* Retrieved from eia.gov: https://www.eia.gov/tools/faqs/faq.php?id=427&t=3

US Environmental Protection Agency (2017, January 24). *Energy and the environment/electricity – customers.* Retrieved from www.epa.gov: https://www.epa.gov/energy/electricity-customers

Woodroof, E. A. (2009). *Green facilities handbook: Simple and profitable strategies for managers.* Lilburn, GA: The Fairmont Press.

World Business Council for Sustainable Development (n.d.). *Making tomorrow's buildings more energy efficient.* Retrieved from www.wbcsd.org: http://www.wbcsd.org/Overview/Resources

World Economic Forum (2017). *Reports.* Retrieved from weforum.org: http://reports.weforum. org/global-energy-architecture-performance-index-report-2016/energy-access-and-security/. Accessed 6 July 2017.

Chapter 17
Implement and Manage the Plan

17.1 Project Management

Project implementation is a discipline of its own, and energy project management shares challenges and methodologies with other technical and people-oriented projects and programs. Managers should adapt their standard project management process as appropriate; good project management skills and techniques will stand champions of the energy and water projects as well as managers in good stead.

17.2 Energy Management

Human energy management, or energy management, is core to the Strategic Framework and should be core to the enterprise itself. Unmanaged projects or programs are likely to be unsuccessful in any field, and when the goal is increased energy effectiveness throughout the organization, the process must be managed throughout the organization. Monitoring and control systems provide the information necessary to accomplish that.

17.3 Awareness and Involvement

Energy projects, like any other initiatives, should not be conceived, designed, or implemented in secret. Although this point has been made at each stage of the Framework, a lack of wide involvement and transparency will become most evident at the implementation stage and can lead to failure.

One frequently cited challenge comes with new building construction. A building designed to the highest energy efficiency standards into which people move

© Springer International Publishing AG, part of Springer Nature 2018

S. McCardell, *Energy Effectiveness*, https://doi.org/10.1007/978-3-319-90255-5_17

without training or understanding of the design principles in that new building is sometimes found to use more energy than planned. The fault is generally not with the design but with the lack of training and awareness. For example:

- Windows may only be operable when the outside air temperature is within certain limits, which building occupants find frustrating.
- Local thermostat controls may be operable only in certain conditions which occupants do not know, leading them to bring in fans or heaters to be comfortable.
- Lighting systems may have been installed with a great amount of flexibility so that occupants can choose direct light, ambient light, or daylighting but without knowing that occupants may bring in their own task and general lighting, defeating the system.

Energy awareness is the first step in understanding and taking action to become more energy effective. In the same way, the first step in ensuring a successful energy project or program is awareness and involvement.

17.4   The Human Factor

If the Framework has been followed, management and staff will have been involved in the information collection and planning stages up to this point and will know about and be supportive of the program. There will also be a designated champion or group leading the project. For the project to be a success, the understanding gained from behavioral economics can then be used to design an implementation process around the way people behave.

Each project will be different, but all should take the following principles into consideration:

- Make the project public and ensure accountability.
- Set up groups or teams to share the work and the successes, because in a connected organization, teams are powerful.
- Focus first on easy, quick wins that can then be cited as examples of success.
- Communicate successes as well as areas for improvement.
- Listen.
- Reward.
- Remember that even small savings can have a large cumulative effect.
- Design around these normal human behaviors:
  - Avoidance of loss is more important than hoping for gain.
  - People are resistant to change but like to be invited and involved.
  - Using guilt as a motivation tool is counterproductive.
  - Relying on technology to save us will not work.
  - People are not generally motivated to conserve, but they are motivated to increase financial measures that have benefit for them.

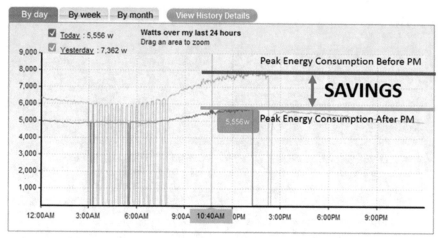

Fig. 17.1 Example of savings identified from power management program. ((Powerhouse Dynamics, 2017) SiteSage "screen capture." Used with permission.)

17.5   Monitoring Devices and Control Systems

Monitoring devices or control systems are the backbone of an energy effectiveness program, although it is possible to design, plan, implement, and manage an energy project manually, with extra staff time and effort. If results are not tracked, energy and water are not likely to be managed, and the changes made are likely to be temporary, forgotten within months.

Monitoring and/or control systems may be installed as part of the energy project, or they may already in place. In either case, good baseline information on energy use and costs should be collected to set the stage for measuring the results of the energy project. The graph in Fig. 17.1, for example, shows the reduced peak energy consumption in a facility after installation of a peak load management device (Fig. 17.1).

With monitors or controls in place, energy and water use are no longer invisible. Project implementation can then focus on managing something that seems more real, and management and staff will all understand that the actions they take (or don't take) will be visible as well. Key performance indicators including these metrics connect to the strategic goals in an integrated approach.

> **Submetering/Monitoring Examples**
> *(1) For example, a commercial office building owner is unsatisfied with the way he allocates utility costs among tenants. There is one meter to the building for each utility, because the owner subdivided the building into 15 units of equal size. Electricity, natural gas, and water/sewer bills are therefore*

(continued)

divided equally among all 15 tenants. The current tenants perceive the utility allocation to be unfair and some are refusing to pay the charges as assessed.

To protect his revenue by reducing tenant turnover, the owner installs simple monitoring systems which serve as submeters and track each tenant's utility use. Once each tenant knows that they are paying for their own measured utility use, they each reduce their usage substantially, and although the owner assesses a 10% administrative fee when billing tenants, each tenant pays less per month on average. The owner increases revenue with the 10% administration fee and reduces tenant turnover because current tenants are pleased with the idea of controlling their own utility costs and decreasing expenses.

(2) In a restaurant, management wants to determine which among the many pieces of equipment should be targeted for energy efficiency improvements first and whether employees are following appropriate shutdown procedures for kitchen equipment. They too decide to install a monitoring system – not as an energy efficiency project but to collect baseline information in preparation for an as-yet-unspecified energy project. The system design and installation are less complicated than feared because the data collection nodes easily attach to wires and pipes and communicate wirelessly. The system is up and running quickly, and almost immediately the following discoveries are made:

- *The oven hood fan is rarely if ever turned off; the constant influx of unconditioned air into the restaurant increases either heating or loads (depending on the season) by about 20%.*
- *During one shift, both hot and cold water at the sink are frequently left to run for up to 10 min at a time.*
- *On another shift, the ovens are usually set to preheat approximately 20 min before they will actually be needed.*

In preparation for the energy project, management begins working with the kitchen staff to better understand the reasons for these results. Once the kitchen staff is able to see the results of their habit patterns, they ask for access to the monitoring system on the computer in the kitchen. Treating "reduced energy and water use" as a recipe they want to perfect, the kitchen staff works to be energy effective in all they do, setting up competitions between shifts and teams. To everyone's surprise, these efforts reduce utility use and cost by over 20% without the need for an energy project.

In this case, the intermediate goal was to determine where energy and water use were highest, so that the main goal – reducing energy and water use – could then be addressed. With supportive and enthusiastic management and staff who took this as a challenge, energy monitors helped the restaurant accomplish the end goal without significant investment.

17.6 Financial Tools

Many of the goals to be addressed through energy effectiveness relate to finances. Just as a good implementation process includes the design and installation of the monitoring/control system, so it must also include the design and adoption of a relevant financial management system or process.

Chapter 9 presented information on financial systems as they relate to energy, including the need to set up a system for reviewing and managing invoices from utility companies connected to the operations and maintenance departments responsible for energy use. Financial analysis methodologies for new projects were also discussed in that chapter.

Once a project is to be implemented, however, additional and different financial tools linked to the key performance indicators developed as part of the SEE Framework should be institutionalized as part of the finance department's standard procedures and reporting structure, where management can rely on them as they do other indicators and metrics.

With standard project management tools, financial management, energy and water management, and a rollout sensitive to the ways in which people behave, programs can be effectively implemented and managed over the long term.

References

ACEEE, American Council for an Energy Efficient Economy. (n.d.). *Programs*. Retrieved from ACEEE.org: http://aceee.org/portal/programs.
Baker, N. e. (2017). *Institutional change Federal Energy Management*. Retrieved from Energy.gov/eere/femp: https://energy.gov/eere/femp/institutional-change-federal-energy-management.
Bertoldi, P. E. (2016, June). *Energy efficiency Volume 09, Issue 03*. Dordrecht: Springer Publishing.
Carayannis, E. e. (2014). *Business model innovation as lever of organizational Sustainability*. New York: Springer Science+Business Media. https://doi.org/10.1007/s10961-013-9330-y.
Diana, F. (2011, 3 9). *The evolving role of business analytics*. Retrieved from Frank Diana's Blog – Our Emerging Future: https://frankdiana.net/2011/03/19/the-evovling-role-of-business-analytics/.
Economist intelligence Unit. (2012). *Energy efficiency and energy savings; a view from the building sector*. London: The Economist Intelligence Unit LTD.
Eggink, J. (2007). *Managing energy costs – a behavioral and non-technical approach*. Lilburn: Fairmont Press.
Energy Information Administration. (n.d.). *Today in energy*. Retrieved from EIA.Gov: https://www.eia.gov/todayinenergy/detail.php?id=18071.
Energy Star. (n.d.). *Use portfolio manager/understand metrics/what energy*. Retrieved from www.energystar.gov/buildings/facility owners and managers: https://www.energystar.gov/buildings/facility-owners-and-managers/existing-buildings/use-portfolio-manager/understand-metrics/what-energy.

Hansen, S. (n.d.). *Making the business case for energy efficiency.* Retrieved from docplayer.net: http://docplayer.net/6344792-Making-the-business-case-for-energy-efficiency-shirley-j-hansen-ph-d.html.

Leonardo Academy. (n.d.). *Free courses and programs on sustainable energy.* Retrieved from Leonardo Academy: www.leonardo-academy.org.

Mourik, R. e. (2015). *What job is energy efficiency hired to do? A look at the propositions and business models selling value instead of energy or efficiency.* Retrieved from IEADSM Leonardo Energy: https://www.youtube.com/watch?v=GGLYp_fHrMs.

New Buildings Institute. (2017, 7 15). *Deep energy retrofits.* Retrieved from New Buildings Institute: http://newbuildings.org/hubs/deep-energy-retrofits.

Powerhouse Dynamics. (2017). *Solutions – Screen captures sent via personal communications.* Retrieved from www.poerhousedynamics.com: https://powerhousedynamics.com/solutions/sitesage/.

Public Technology Inc./US green building council. (1996). *Sustainable building technical manual.* Washington, DC: Public Technology, Inc.

Russell, C. C. (2010). *Managing energy from the top down; connecting industrial energy efficiency to business performance.* Lilburn: The Fairmont Press Inc/CRC Press.

Shields, C. (2010). *Renewable energy facts and fantasies.* New York: Clean Energy Press.

SustainAbility. (2014). 20 business model innovations for Sustainability. SustainAbility.

The Shift Project. (2012). *Top 20-Capacity Chart.* Retrieved from htttp://www.tsp.org the shift project: http://www.tsp-data-portal.org/TOP-20-Capacity#tspQvChart. Accessed 2/17/2017.

US Department of Energy Energy Efficiency and Renewable Energy. (2011). *A guide to Energy Audits PNNL-20956.* Pacific Northwest National Laboratory. Retrieved from http://www.pnnl.gov/main/publications/external/technical_reports/pnnl-20956.pdf.

US Energy Information Agency. (n.d.). *faqs.* Retrieved from eia.gov: https://www.eia.gov/tools/faqs/faq.php?id=427&t=3.

US Environmental Protection Agency. (2017, 1 24). *Energy and the environment/electricity – Customers.* Retrieved from www.epa.gov: https://www.epa.gov/energy/electricity-customers.

Woodroof, E. A. (2009). *Green facilities handbook – simple and profitable strategies for managers.* Lilburn: The Fairmont Press.

World Business Council for Sustainable Development. (n.d.). *Making tomorrow's buildings more energy efficient.* Retrieved from www.wbcsd.org: http://www.wbcsd.org/Overview/Resources.

World Economic Forum. (2017). *Reports.* Retrieved from weforum.org: http://reports.weforum.org/global-energy-architecture-performance-index-report-2016/energy-access-and-security/. Accessed 7/6/2017.

Chapter 18
The Feedback Loop

18.1 Evaluation, Recognition, and Adaptation

There is a saying with energy projects that the question is less one of "What can be achieved?" than of "What *will* be achieved? This refers to the gap between energy projects that are technically and economically feasible, and those that are implemented.

It might be added that there is also a gap between those projects which are implemented and those with results which are documented and can be demonstrated. With automated real-time measurement capability and information displays that can also capture historical data, that gap is shrinking; evaluation is becoming much easier. As part of the Strategic Framework, the feedback portion of the process includes:

- Evaluation
- Recognition
- Adaptation

Evaluation is based on comparing the actual results from an energy project or program to those which had been anticipated or hoped for. Using the information already collected by monitoring or control systems, the evaluation compares the current situation to the base case, informed by the purpose, goals, key performance indicators or KPIs, and other decisions made as part of the SEE Framework. The comparison will be different for each project and each enterprise but might include:

- Progress towards meeting competitive targets
- Progress towards achieving best practices or industry standards
- Progress towards reducing utility costs, usage, peak/demand charges, or other related goals
- Progress towards meeting related strategic goals
- Progress towards installing a large project
- Progress towards improving internal systems, as that progress can be measured
- Progress towards increased staff awareness and involvement

© Springer International Publishing AG, part of Springer Nature 2018
S. McCardell, *Energy Effectiveness*, https://doi.org/10.1007/978-3-319-90255-5_18

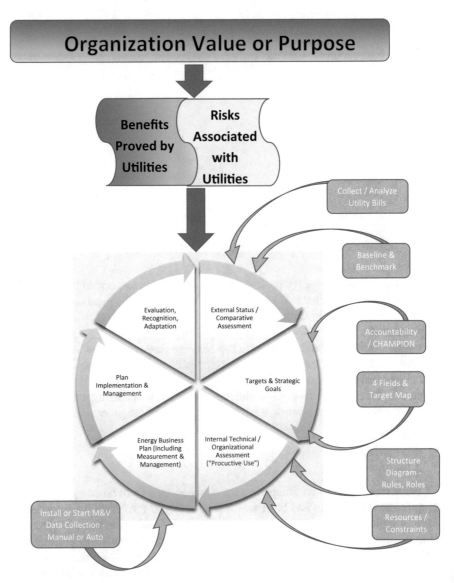

Fig. 18.1 The SEE Framework: Detail (Copyright 2016–2018 Current-C Energy Systems, Inc. Used with Permission.)

(There are likely to be other goals which were defined for each project, as well.) Once the evaluation is complete, it should be widely shared within the organization and possibly outside it as well.

Recognition consists of acknowledging what has been accomplished, what has not been accomplished, and the people or teams who have had a major role in the process and its success. The last point is particularly important, because positive reinforcement leads to organizational change and, according to the behavior change tenets presented in Chap. 9, recognition is one of the most important requirements for success.

Adaptation links the current program to the next round of efforts designed to increase energy effectiveness. If the current initiative is a one-time project, the successes and opportunities for improvement should be captured for the next project – even if it is not scheduled immediately. If the current initiative is an ongoing program, then the adaptation can be immediate, building on successes and adjusting to compensate for elements which did not work as expected. Both projects and programs could adapt by moving in several different directions, including:

- Continue as before.
- Drop the program.
- Increase/decrease support for the program.
- Involve more or different people in the program or project.
- Continue to the next project planned.
- Substantially adjust minor or significant elements of the initiative.
- Migrate the initiative to another division or facility.

The adaptation phase connects this phase of the project or program to the next. Once the appropriate direction has been determined, the feedback is built into a new cycle. On the assumption that the purpose of the organization remains the same and that there is little change in the risks and benefits associated with the utilities, the cycle continues with the segments subject to energy and water management. The next step is therefore the analysis phase and the external assessment. With each successive iteration, the organization will become more energy effective.

In summary, the Feedback loop phase considers in detail all previous phases in the SEE Framework. For reference, those detailed phases are shown in Fig. 18.1.

References

ACEEE, American Council for an Energy Efficient Economy. (n.d.). *Programs*. Retrieved from ACEEE.org: http://aceee.org/portal/programs.

Baker, N. e. (2017). *Institutional change Federal Energy Management*. Retrieved from Energy.gov/ eere/femp: https://energy.gov/eere/femp/institutional-change-federal-energy-management.

Bertoldi, P. E. (2016, June). *Energy efficiency Volume 09, Issue 03*. Dordrecht: Springer Publishing.

Carayannis, E. e. (2014). *Business model innovation as lever of organizational Sustainability*. New York: Springer Science+Business Media. https://doi.org/10.1007/s10961-013-9330-y.

Diana, F. (2011, 3 9). *The evolving role of business analytics*. Retrieved from Frank Diana's Blog – Our Emerging Future: https://frankdiana.net/2011/03/19/the-evovling-role-of-business-analytics/.

Economist intelligence Unit. (2012). *Energy efficiency and energy savings; a view from the building sector*. London: The Economist Intelligence Unit LTD.

Eggink, J. (2007). *Managing energy costs – a behavioral and non-technical approach*. Lilburn: Fairmont Press.

Energy Information Administration. (n.d.). *Today in energy*. Retrieved from EIA.Gov: https://www.eia.gov/todayinenergy/detail.php?id=18071.

Energy Star. (n.d.). *Use portfolio manager/understand metrics/what energy*. Retrieved from www.energystar.gov/buildings/facility owners and managers: https://www.energystar.gov/buildings/

facility-owners-and-managers/existing-buildings/use-portfolio-manager/understand-metrics/
what-energy.

Hansen, S. (n.d.). *Making the business case for energy efficiency.* Retrieved from docplayer.net:
http://docplayer.net/6344792-Making-the-business-case-for-energy-efficiency-shirley-j-han-
sen-ph-d.html.

Leonardo Academy. (n.d.). *Free courses and programs on sustainable energy.* Retrieved from
Leonardo Academy: www.leonardo-academy.org.

Mourik, R. e. (2015). *What job is Energy Efficiency hired to do? A look at the propositions
and business models selling value instead of energy or efficiency.* Retrieved from IEADSM
Leonardo Energy: https://www.youtube.com/watch?v=GGLYp_fHrMs.

New Buildings Institute. (2017, 7 15). *Deep energy retrofits.* Retrieved from New Buildings
Institute: http://newbuildings.org/hubs/deep-energy-retrofits.

Powerhouse Dynamics. (2017). *Solutions – Screen captures sent via personal communications.*
Retrieved from www.poerhousedynamics.com: https://powerhousedynamics.com/solutions/
sitesage/.

Public Technology Inc./US green building council. (1996). *Sustainable Building Technical
Manual.* Washington, DC: Public Technology, Inc.

Russell, C. C. (2010). *Managing energy from the top down; connecting industrial energy efficiency
to business performance.* Lilburn: The Fairmont Press Inc/CRC Press.

Shields, C. (2010). *Renewable energy facts and fantasies.* New York: Clean Energy Press.

SustainAbility. (2014). 20 business model innovations for Sustainability. SustainAbility.

The Shift Project. (2012). *Top 20-Capacity Chart.* Retrieved from htttp://www.tsp.org the shift
project: http://www.tsp-data-portal.org/TOP-20-Capacity#tspQvChart. Accessed 2/17/2017.

US Department of Energy Energy Efficiency and Renewable Energy. (2011). *A guide to Energy
Audits PNNL-20956.* Pacific northwest national laboratory. Retrieved from http://www.pnnl.
gov/main/publications/external/technical_reports/pnnl-20956.pdf.

US Energy Information Agency. (n.d.). *faqs.* Retrieved from eia.gov: https://www.eia.gov/tools/
faqs/faq.php?id=427&t=3.

US Environmental Protection Agency. (2017, 1 24). *Energy and the environment/electricity –
Customers.* Retrieved from www.epa.gov: https://www.epa.gov/energy/electricity-customers.

Woodroof, E. A. (2009). *Green facilities handbook- simple and profitable strategies for managers.*
Lilburn: The Fairmont Press.

World Business Council for Sustainable Development. (n.d.). *Making tomorrow's buildings more
energy efficient.* Retrieved from www.wbcsd.org: http://www.wbcsd.org/Overview/Resources.

World Economic Forum. (2017). *Reports.* Retrieved from weforum.org: http://reports.weforum.
org/global-energy-architecture-performance-index-report-2016/energy-access-and-security/.
Accessed 7/6/2017.

Part VI
Introducing and Using the Strategic Energy Effectiveness Framework (SEE): Putting It All Together

Chapter 19
The Strategic Energy Effectiveness Framework

19.1 SEE Framework Summary

Previous sections of this book have developed the elements of SEE Framework which can be used to find, evaluate, plan, and act upon energy and water opportunities. This chapter builds upon the descriptions and background presented for each segment and presents the Framework as a whole.

The SEE Framework is flexible and able to account for differences in both inputs and expected results, appropriate for different organizations. It also has the potential to be rigorous, assisting managers to understand energy and water inputs and improve outputs. Most importantly, the Framework is designed to be actionable.

The Framework should be addressed in steps, all of which fall into five general stages with sub-segments:

1. Strategy development

   – Clarify the business value or purpose.
   – Clarify the relationship of utilities to that purpose and the related utility benefits and risks.

2. Analysis

   – Consider the external environment and context, including a comparison to similar organizations.
   – In light of the external environment, set goals and targets which take into consideration the role of utilities in supporting or delivering the business purpose.

© Springer International Publishing AG, part of Springer Nature 2018
S. McCardell, *Energy Effectiveness*, https://doi.org/10.1007/978-3-319-90255-5_19

- Perform an internal assessment, identifying and understanding the four elements of usage – building, people, process, and equipment/environment; in each area, identify areas of waste and opportunity (including co-benefits).

3. Planning

   - Develop a business energy plan including a strategy for measurement and evaluation/verification.

4. Implementation

   - Implement the plan, taking action first on small items where success can be easily achieved.

5. Feedback

   - Measure, evaluate, adjust, and continue, going back to step 2.

19.2   The Strategic Energy Effectiveness Framework in Practice

SEE frames a process of developing energy effectiveness, as shown in Fig. 19.1, which provides chapter references and more detail on each segment in the Framework. They are each addressed in turn in the balance of this summary chapter.

19.2.1   Strategy

| Value or Purpose | • Relation of Value or Purpose to Utilities
• Impact of Utililties on Value or Purpose
• (Chapter 10) |

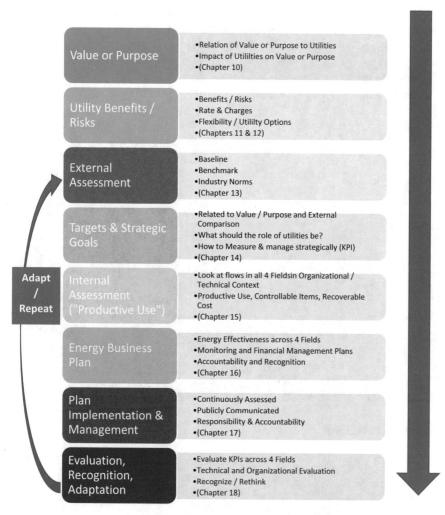

DEVELOPING ENERGY EFFECTIVENESS

Fig. 19.1 Developing energy effectiveness (Copyright 2016–2018 Current-C Energy Systems, Inc. Used with permission)

Clarify the Business Purpose or Value and the Relationship of Utilities to that Purpose

The business purpose and value of any type of entity (including nonprofits, government entities, or mining concerns) are important to consider during the process of addressing energy and water issues, because those resources should enable what the organization is working to achieve. Whatever it is, utilities should support that purpose, and if there are energy or utility-related issues which impede it, they should be addressed; if there are energy or utility-related opportunities that enhance that purpose, they should be taken.

Purpose Connected to Energy and Water: Examples

Manufacturing company A is a contract manufacturer providing products for multiple long-term customers. Their purpose is to keep those customers satisfied by providing excellent quality and flexibility at a good price. Their manufacturing equipment is all electrical, and they run 2–3 shifts depending on customer orders. Electricity is thus the single energy source, and any changes in the way it is used must not compromise either quality or flexibility. As they begin planning for a project to reduce electrical use, management assembles a team with strong technical/manufacturing and customer support representation to develop options that serve their purpose while also reducing costs.

Service company B is nonprofit which trains and places individuals who have been downsized or are otherwise looking for new positions or careers. They are located in an old leased building in a marginal downtown area. Their purpose is to provide a safe, comfortable location for clients who are feeling unsure of themselves and their future, increasing economic activity in the surrounding area. As they begin planning for a project to reduce utility use, they note that some of their clients are concerned that any potential changes not impact their feelings of safety and comfort, while others have relevant expertise in engineering, wiring, plumbing, cleaning, or construction. Clients are asked if they are interested in volunteering to help develop options for reducing energy costs and improving the building, and several do so – adding both concerns and expertise to the team and serving the Purpose while potentially reducing costs.

Utility Benefits/Risks

Utility Benefits / Risks	• Benefits / Risks • Rate & Charges • Flexibility / Utililty Options • (Chapters 11 & 12)

Understand Energy Inputs

Much like the child who grows up in the city and answers the question "Where does milk come from?" with "The grocery store!," most managers would know that their power, natural gas, and water arrive in distribution pipes or lines, but not what happens before then. Becoming aware of the risks and benefits associated with the utilities on which an enterprise relies enables a deeper understanding of their importance. Some utilities may be monopoly providers, while others may operate in a competitive market. Knowing what utilities provide, how that is generated or sourced and transmitted, and the components of the pricing structure are important pieces of information. And managers must understand what they do with the energy and water they receive. Options for alternative energy sources, either locally or delivered by utilities, might also be relevant.

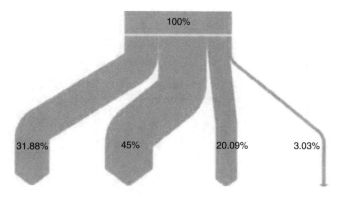

Fig. 19.2   Example of a simple Sankey diagram

Conceptually, one way to look at this issue would be to construct a simple Sankey diagram showing the energy or water coming in, if possible including the sources of generation, and the uses/waste inside the facility.

"The typical manufacturing facility must inflate its energy procurement budget by an additional two-thirds to account for the energy that it will eventually waste."[1] Waste can be expensive.

19.2.2   Analysis

External Assessment: Consider the External Environment and Context

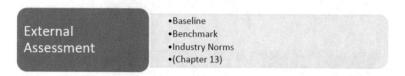

In the same way that organizations considering a new product or service conduct a competitive analysis, those who are considering the effectiveness of their utilities should compare their operations to those of other similar organizations. There are many ways to do that, including Energy Star in the United States or industry average statistics from organizations such as the International Energy Administration, or IEA. Other elements of this external analysis might include researching the sources of energy currently provided (e.g., hydropower is generally considered a renewable resource, but availability is subject to increasing drought).

Both company A and service company B in the Comparative Assessment Examples validated the level of their energy use by comparing it to other similar buildings. There are two elements of understanding energy inputs – first is seeing what the charges are, including subcomponents; the second is comparing usage to

---

[1] Managing Energy Russell

Comparative Assessment: Examples (Continued)

Manufacturing company A relies on electricity for 100% of its energy input and (without a cafeteria, laundry, or showers) uses little water. Since they adjust their production line based on customer orders, their production process is variable and electricity use inconsistent. After reviewing their utility bills and determining that they have high demand charges during certain periods, they decide to do a comparative analysis. Referencing the Energy Star database after inputting their information into Portfolio Manager[2], they determine that, in comparison to other similar manufacturing operations, their utility usage is about 1.5 times the average, but their utility costs are over twice the average.

Service company B, the nonprofit in the leased downtown facility, has its own natural gas, electricity, and water meters. They note that their water use (primarily for the cafeteria) has been inconsistent through the past 2 years, without any discernable seasonal pattern. Natural gas is used for heating and cooking and turns out to be lower than the average shown in Energy Star, a surprising result given the age of the building. Electricity use is primarily for lighting and office equipment and is comparatively high.

Fuel Mix Comparison

This chart compares fuel mix (%) of sources used to generate electricity in your region to the fuel mix (%) for the entire United States.

Fig. 19.3 Fuel mix comparison example (https://www.epa.gov/energy/power-profiler)

that of other similar facilities; Energy Star is one common resource to use for that comparison in the United States; others can be found in Chap. 21.

The US EPA's Energy Profiler provides a simple way to determine the fuel mix for any facility in the United States, by zip code with additional explanatory information as shown in Fig. 19.3.

---

[2] Energy Star portfolio manager

Targets and Strategic Goals Should Be SMART

Targets & Strategic Goals	•Related to Value / Purpose and External Comparison •What should the role of utilities be? •How to Measure & manage strategically (KPI) •(Chapter 14)

The SMART goals should be ones which lead to achieving the purpose or value, and they can be used to screen out options which do not meet that purpose, taking into consideration the role that utilities play in the organization.

SMART is an acronym for goals which are well defined, actionable, and measurable. The exact terms used to form the acronym differ from expert to expert, but these convey the basic idea:

S = Strategic
M = Measurable
A = Actionable
R = Reasonable
T = Time-bounded

A SMART goal format is provided in Chap. 20, and of course the terms can be adapted as appropriate for the project. For example, S might be shared, or A might be achievable.

SMART Goals as a Constraint: Example

A hospital in a remote area with unreliable power has a gas generator which is past its end of life and beginning to stutter. Since the hospital's business purpose (and many hospital regulations) require that the facility be open 24 h a day, every day, using backup power if necessary. The hospital is financially constrained, like most small rural hospitals, but knows that in developing options one choice it does NOT have is ignoring the generator issue until it finally does not work. The SMART goals that hospital management develops must recognize that reality.

Internal Assessment: Productive Use

Internal Assessment ("Productive Use")	•Look at flows in all 4 Fieldsin Organizational / Technical Context •Productive Use, Controllable Items, Recoverable Cost •(Chapter 15)

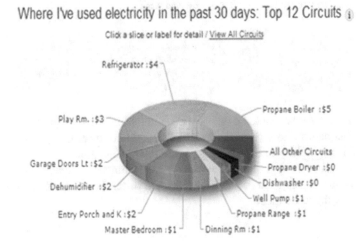

Fig. 19.4 Example of monitoring summary by circuit. ((Powerhouse Dynamics, 2017) EnergySage. Used with permission)

The steps involved in performing an internal assessment include identifying and understanding the elements of usage in the Four Fields – building, people, process/procedures/equipment, and environment – and then determining where the opportunities are.

This assessment could take the form of adding detail to the very general Sankey diagram shown in Fig. 19.2. Sankey diagrams can be developed manually or using online tools; at the time of writing, one option is http://sankeymatic.com/. Where energy monitors or controls are installed, the graphic displays of energy use by equipment, building, or category (such as lighting) can be used directly. (Fig. 19.4 shows usage in a home, highlighting the fact that energy monitors have a place even in small buildings and enterprises). Manual estimates basically consist of counting every piece of equipment in the facility and calculating or estimating its energy use; a format for estimating these is shown in Chap. 20.

Using the SEE Framework, the internal assessment is completed after the goals are determined to economize on the effort required. If all the goals relate to reducing water use, for example, the time spent collecting information on natural gas and electricity usage is of marginal benefit. This is different from the standard approach where a consultant or energy service company conducts an energy assessment or energy audit of all systems and uses as a first step.

Identify Areas of Waste and Opportunity (Including Co-Benefits)

Within the context of the internal assessment and based on the information collected during that process, the next step is to identify areas where waste can be recaptured; a checklist of normal areas to investigate challenges and opportunities is shown in Chap. 20, and of course it should be adapted as appropriate. One useful technique is to map out the sources of waste and ways in which that waste might be recaptured for other useful processes.

Another useful technique for segregating the waste recapture opportunity from the productive work the energy or water does is to divide the budget for electricity, natural gas, and water along the lines normally used by the enterprise (such as by department or by product line) and then divide each of those into two line items representing the value of the energy required to perform useful work and the value of the energy that will be wasted,[3] as shown in the following abbreviated example:

Total natural gas cost: $500,000
   Space heating:
     Productive use: $75,000
     Waste: $25,000
   Production equipment:
     Productive use: $250,000
     Waste: $250,000

Many energy projects are not undertaken because managers "have no money". The analysis above shows that they DO, but it is earmarked to spend on future purchases of energy that will be wasted.

19.2.3   Planning

The Energy Business Plan

Energy Business Plan	• Energy Effectiveness across 4 Fields • Monitoring and Financial Management Plans • Accountability and Recognition • (Chapter 16)

The next step is to develop a business energy plan including a strategy for measurement and evaluation/verification.

The industry is split regarding the advisability of starting with a small project or doing a complete energy efficiency upgrade. The whole building or deep projects do save significantly more (often with substantial investments), but for organizations which are just developing an energy-effective culture, it is more important to begin a program somewhere, measure results, and then continue the program. In such cases, the best approach is to start with easy wins that include co-benefits. These might include replacing aging equipment with more efficient models, upgrading a leaky roof by improving insulation and simultaneously installing a solar system and so on. Quick wins might also be projects for which there is widespread popular support within the organization or where grants and incentives apply; one example might be using municipal incentives for replacing lawns with landscaping that requires less watering and provides shaded seating areas.

---

[3] Russell p 58

Whether the project is large or small, it should begin with an Energy Business Plan complete with SMART goals, costs, and benefits. The plan should use a methodology to which management of the company is accustomed including the appropriate financial criteria, success measures, and evaluation methodologies.

Within the energy business plan, a key element is to focus on the people, including the commitment of leadership to the project, involvement throughout the organization, and the designation of a champion individual or group.

Energy business plans are important for one-time project investments in equipment as well as longer-term programs designed to change behavior; since people are involved in either case, part of the plan should include structures and actions intended to keep the program interesting. Documentation, evaluation, and other metrics should also be planned.

The financial metrics used in the plan should be those with which managers are familiar such as life cycle costing, return on investment, or others.

19.2.4   Implementation

Plan Implementation & Management	•Continuously Assessed •Publicly Communicated •Responsibility & Accountability •(Chapter 17)

The next step is of course to implement the plan, including the measurement methodology which might consist of new equipment to monitor usage or the development/adaptation of current reports. This is also the beginning of the energy management process with its core related to the Strategic Energy Effectiveness Framework and and leading to an ongoing process devoted to increasing energy effectiveness.

19.2.5   Feedback

Measure, Evaluate, Adjust, and Continue

Evaluation, Recognition, Adaptation	•Evaluate KPIs across 4 Fields •Technical and Organizational Evaluation •Recognize / Rethink •(Chapter 18)

As with most efforts, a feedback loop is critical, with continued focus on making the program better, stronger, and more energy effective. In some ways, it is like cleaning a house or applying for ISO certification. Each cycle should show improvement, but at the end of each cycle, the process begins again.

In the energy field, that would be considered an enduring opportunity.

References

ACEEE, American Council for an Energy Efficient Economy. (n.d.). *Programs*. Retrieved from ACEEE.org: http://aceee.org/portal/programs.

AmericanCouncilforanEnergyEfficientEconomy.(2015,2).*aceee.org/blog*.RetrievedfromACEEE: http://aceee.org/blog/2015/02/why-we-don%E2%80%99t-have-choose-between-ener.

APC. (n.d.). *tag/efficiency/page2*. Retrieved from www.sankey-diagrams.com: http://www.sankey-diagrams.com/tag/efficiency/page/2/.

Arch Tool Box. (n.d.). *Energy use intensity*. Retrieved from Archtoolbox.com, Architect's Technical Reference: https://www.archtoolbox.com/sustainability/energy-use-intensity.html.

Argus Technologies. (2014–2017). *energiemonitoring*. Retrieved from argustech.be: https://argustech.be/en/energiemonitoring/.

Cantor, J. (2017). *Energy monitoring*. Retrieved from heatpumps.co.uk and heatpumps.co.uk: http://heatpumps.co.uk/technical/energy-monitoring/.

Carayannis, E. e. (2014). *Business model innovation as lever of organizational sustainability*. New York: Springer Science+Business Media. https://doi.org/10.1007/s10961-013-9330-y.

Cooremans, C. (2012). Investment in energy efficiency: Do the characteristics of investments matter? *Springer Science_Business Media B.V.2012*, pp. 497–518.

Diana, F. (2011, 3 9). *The evolving role of business analytics*. Retrieved from Frank Diana's Blog – Our Emerging Future: https://frankdiana.net/2011/03/19/the-evovling-role-of-business-analytics/.

Economist intelligence Unit. (2012). *Energy efficiency and energy savings; a view from the building sector*. London: The Economist Intelligence Unit LTD.

Energy Information Administration. (n.d.). *Today in energy*. Retrieved from EIA.Gov: https://www.eia.gov/todayinenergy/detail.php?id=18071.

Energy Star. (n.d.). *Use portfolio manager/understand metrics/what energy*. Retrieved from www.energystar.gov/buildings/facility owners and managers: https://www.energystar.gov/buildings/facility-owners-and-managers/existing-buildings/use-portfolio-manager/understand-metrics/what-energy.

Fuller, S. (1994, Updated 2016). *Resources/life cycle cost analysis LCCA*. Retrieved from Sustainable Building Technical Manual/Joseph J Romm, Lean and Clean Management: https://www.wbdg.org/resources/life-cycle-cost-analysis-lcca.

Green Rhino Energy. (n.d.). *The energy value chain*. Retrieved from Greenrhinoenergy.com: http://www.greenrhinoenergy.com/renewable/context/energy_value_chain.php.

Hansen, S. (2002). *Manual for intelligent energy services*. Lilburn: Fairmont Press.

Hansen, S. (n.d.). *Making the business case for energy efficiency*. Retrieved from docplayer.net: http://docplayer.net/6344792-Making-the-business-case-for-energy-efficiency-shirley-j-hansen-ph-d.html.

International Energy Agency. (2015). *Capturing the multiple benefits of energy efficiency.html*. Retrieved from www.iea.org: (Jewell, 2014).

Lawrence Livermore National Laboratory. (2016). *Flowcharts*. Retrieved from llnl.gov Lawrence Livermore National Library: https://flowcharts.llnl.gov/content/assets/images/energy/us/Energy_US_2015.png.

Leonardo Academy. (n.d.). *Free courses and programs on sustainable energy*. Retrieved from Leonardo Academy: www.leonardo-academy.org.

Microsoft. (2016). *Data analytics and smart buildings increase comfort and energy efficiency*. Retrieved from www.microsoft.com: https://www.microsoft.com/itshowcase/Article/Content/845/Data-analytics-and-smart-buildings-increase-comfort-and-energy-efficiency.

Mourik, R. e. (2015). *What job is energy efficiency hired to do? A look at the propositions and business models selling value instead of energy or evviciency*. Retrieved from IEADSM Leonardo Energy: https://www.youtube.com/watch?v=GGLYp_fHrMs.

Powerhouse Dynamics. (2017). *Solutions – Screen captures sent via personal communications*. Retrieved from www.poerhousedynamics.com: https://powerhousedynamics.com/solutions/sitesage/.

Public Technology Inc./US Green Building Council. (1996). *Sustainable building technical manual*. Washington, DC: Public Technology, Inc.

Russell, C. C. (2010). *Managing energy from the top down; Connecting industrial energy efficiency to business performance*. Lilburn: The Fairmont Press Inc/CRC Press.

Studney, C. (2012, 10 4). *How to measure the ROI of LEED*. Retrieved from www.greenbiz.com/blog: https://www.greenbiz.com/blog/2012/10/04/how-measure-roi-leed.

SustainAbility. (2014). *20 Business Model Innovations for Sustainability*. SustainAbility.

The Shift Project. (2012). *Top 20-Capacity Chart*. Retrieved from htttp://www.tsp.org The Shift Project: http://www.tsp-data-portal.org/TOP-20-Capacity#tspQvChart. Accessed 2/17/2017.

US Department of Energy Energy Efficiency and Renewable Energy. (2011). *A guide to Energy Audits PNNL-20956*. Pacific Northwest National Laboratory. Retrieved from http://www.pnnl.gov/main/publications/external/technical_reports/pnnl-20956.pdf.

US Energy Information Agency. (n.d.). *faqs*. Retrieved from eia.gov: https://www.eia.gov/tools/faqs/faq.php?id=427&t=3.

US Environmental Protection Agency. (2017, 1 24). *Energy and the environment/electricity – Customers*. Retrieved from www.epa.gov: https://www.epa.gov/energy/electricity-customers.

Woodroof, E. A. (2009). *Green facilities handbook- simple and profitable strategies for managers*. Lilburn: The Fairmont Press.

World Business Council for Sustainable Development. (n.d.). *Making tomorrow's buildings more energy efficient*. Retrieved from www.wbcsd.org: http://www.wbcsd.org/Overview/Resources.

World Economic Forum. (2016). *Energy access and security*. World Economic Forum.

World Economic Forum. (2017). *Reports*. Retrieved from weforum.org: http://reports.weforum.org/global-energy-architecture-performance-index-report-2016/energy-access-and-security/. Accessed 7/6/2017.

Chapter 20
Framework Checklists

20.1 Strategic Energy Effectiveness Framework (SEE): Detail[1]

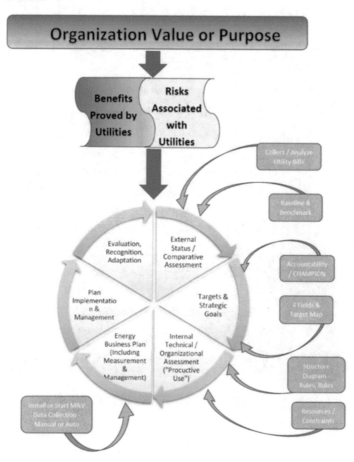

Note: All checklists and forms copyright 2015–2017 Current C Energy Systems, Inc., unless otherwise noted.

[1] Copyright 2016–2018 Current C Energy Systems, Inc. Used with permission

20.2   Developing Energy Effectiveness

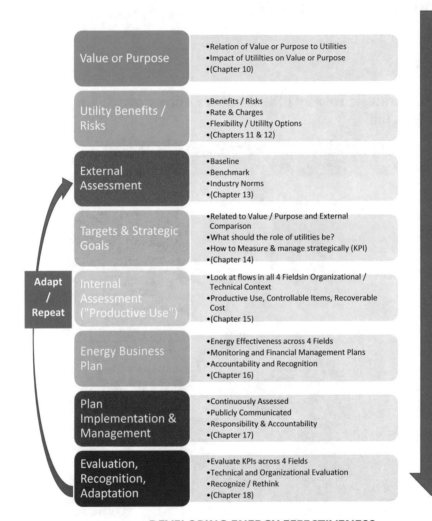

Value or Purpose	• Relation of Value or Purpose to Utilities • Impact of Utililties on Value or Purpose • (Chapter 10)
Utility Benefits / Risks	• Benefits / Risks • Rate & Charges • Flexibility / Utililty Options • (Chapters 11 & 12)
External Assessment	• Baseline • Benchmark • Industry Norms • (Chapter 13)
Targets & Strategic Goals	• Related to Value / Purpose and External Comparison • What should the role of utilities be? • How to Measure & manage strategically (KPI) • (Chapter 14)
Internal Assessment ("Productive Use")	• Look at flows in all 4 Fieldsin Organizational / Technical Context • Productive Use, Controllable Items, Recoverable Cost • (Chapter 15)
Energy Business Plan	• Energy Effectiveness across 4 Fields • Monitoring and Financial Management Plans • Accountability and Recognition • (Chapter 16)
Plan Implementation & Management	• Continuously Assessed • Publicly Communicated • Responsibility & Accountability • (Chapter 17)
Evaluation, Recognition, Adaptation	• Evaluate KPIs across 4 Fields • Technical and Organizational Evaluation • Recognize / Rethink • (Chapter 18)

Adapt / Repeat

DEVELOPING ENERGY EFFECTIVENESS

20.3   Examples of Opportunities by Field and Type

Field	Area	Incremental opportunity/low cost or no cost	System/investment required
Process	Load shifting	Reschedule systems and equipment to reduce demand fees within day, season, etc.	
Building	Indoor lighting	Turn off when not used; use daylight when possible; use occupancy sensors	Make sure level matches required use; reduce overlighting, relamp; retrofit lighting; install lighting control system. Use LEDs wherever possible
Building	Building envelope	Replace damaged insulation, install awnings over windows, plug holes around vents, etc.	Insulate walls, roof/ceiling, and floor
Equipment	HVAC	Make sure sensors work and control set points set to correct temp, optimize on/off controls to match HVAC use with demand, turn off when not in use, and make sure cooling and heating not in conflict	Retro-commission HVAC system so that it performs as it was designed and is appropriate for current building layout
People	Team	Use enthusiasm of management, champion, and team for energy conservation; have contests; reward success; provide time and minimal funds	Much greater success with monitoring and/or control systems installed
Process	Compressed air	Fix leaks, match/reduce pressure to what the equipment needs	Make sure connected to use and not leaking; make sure all filters cleaned/replaced frequently
Process	Boiler/steam system	Fix steam trap and pipework leaks, insulate steam and hot water lines, and clean boiler to reduce fouling and scale	Ask specialist to check boiler flue gases for complete combustion/optimal excess air ratio
Process	Refrigeration	Make sure system is not blocked/fouled, e.g., build up on evaporators; keep cold room doors closed when not in use	Inspect and clean condensers and evaporators regularly
Process/ equipment	Motors/drives/pumps	Turn equipment off when not being used, and make sure motors and pumps are not oversized	Have clear decision-making process for new/rewinding decision; install VSDs/VFDs and monitoring systems

(continued)

Field	Area	Incremental opportunity/low cost or no cost	System/investment required
Processes	Mtce processes	Include ongoing inspections of energy using systems especially compressed air, steam traps, HVAC, lighting, and building; make sure system for reporting issues works	Set up "group replacement" programs for lights, equipment, computers, etc.
Processes/ human factors	Energy management system/monitors	If in place, make real-time data available widely; confirm all specifications operating as they should. Understand components and drivers	Install more detailed EMS or monitoring systems as possible
Processes/ equipment	Plumbing, pipework	Insulate all plumbing which is cooler/warmer than surroundings; check frequently for leaks	Replace pipes as necessary
Equipment	Computers, printers, coffee makers, office equipment, and other plug loads	Turn off when not being used especially at the end of the day; install auto power strips with occupancy sensors or timers	Install systems which shut down office equipment and other plug loads
Equipment/ human factors	Individual plug loads	Keep personal individual plug loads out of office – Refrigerators, microwaves, heaters, incandescent desk lamps, water coolers	Install large shared refrigerator and shared water cooler; retro-commission HVAC system so it works properly
Equipment	Printers/copiers	Turn off when not in use; install power strips which do so automatically	
Procedures	Procurement policies	Develop procurement policies for new purchases that require Energy Star equipment, consideration of amount of consumables used, etc.	
Environment	Outdoor lighting	If lights are on timers, adjust throughout the year	Install LEDs; install light sensors for adjustable timing during the year
Procedures	Cleaning	Consider moving to daylight to save on lighting and higher hourly fees, if it can be done to avoid bothering occupants	
Procedures/ human factors	Metering	Track daily/real-time consumption and costs, analyze, and take action	
Building	Lighting switches	Make sure they are easy to access	Set up smaller lighting zones; install occupancy sensors or full control systems

(continued)

Field	Area	Incremental opportunity/low cost or no cost	System/investment required
Building	Exit signs	Replace all exit signs with LEDs to improve safety and reduce energy use	
Building/ human factors	Painting	Paint in light, bright colors to reduce lighting need, and improve the look of the facility as well as occupant outlook	
Equipment/ human factors	HVAC	Set reasonable set points, summer and winter	
Building	Openings	Keep doors and windows closed as much as possible when HVAC systems are running. Alternatively, turn the HVAC system off, and use natural cooling	
Processes	Manufacturing line	Consider adjusting work schedules to avoid peak demand periods if possible or stagger start times	
Procedure	Utility rate charge	Once utility use is understood, analyze current utility rates, and determine if it should be negotiated or not	
Equipment	Fans	Adjust fan speeds where possible, and adjust airflow	Anti-stratification fans for high areas to keep heated or cooled air where people are
Environment	Building exterior	Use light colors outside in hot climates to reflect heat and reduce a/c charges	
Environment	Parking area	Use light-colored pavement for parking area, and make it permeable so stormwater and melting ice can percolate through	
Building/ human factors	Envelope/openings	Use shading and blinds to reduce glare and heat through windows	
Equipment	HVAC	Seal heating and cooling ductwork, and maintain frequently	Install variable frequency drives on fan motors
People	Awareness	Conduct energy and water awareness training, and assign related tasks with rewards for savings – Make it fun	

(continued)

Field	Area	Incremental opportunity/low cost or no cost	System/investment required
Procedures, people	Metrics	Determine performance metrics and key performance indicators such as amount of energy per finished product, amount of water per hour, etc. as appropriate	
Processes	Air compressor	Ensure air compressor air intake is outside or in coolest location; compression takes less energy; discharge to areas requiring heat	
Equipment	Equipment	Use all equipment at high end of capacity where it is more efficient	Replace larger equipment with smaller HP if it cannot be run efficiently
Equipment	Refrigeration	Keep refrigeration units full; they retain cold better, so use less electricity	
Environment	Irrigation systems	Conduct frequent maintenance to prevent and catch/repair leaks	Install or use moisture sensors to water only where it is required
Environment	Shading	Install landscaping (vines, deciduous trees, trellises, overhangs) to shade windows and for protection from sun/wind	
Processes	Water	Install low-flow faucets and showerheads, using high-velocity low-volume models	Consider installing recirculating pumps at the end of long plumbing runs for hot water
Processes	Cooking	Turn kitchen hood exhausts off whenever not necessary, to reduce electricity use and heating/cooling needs	
Processes	Swimming pools	Place cover over pool when not in use – Manual cover	Install automatic pool cover to be used before and after hours, saving from reduced evaporation, heating needs, and chemical costs
Processes	Laundry facilities		Install heat exchanger to capture heat from wastewater to preheat wash water

20.4   Paradigm Shift Strategies to Bring Energy/Water Under Control and Become Energy Effective

Objective: paradigm shift	Energy/ water as a controllable / manageable cost	Energy effectiveness is the end state
Current state	Changes needed	Desired end state

20.5   Leverage Points for Waste Recapture Potential

TOTAL POTENTIAL RECAPTURE

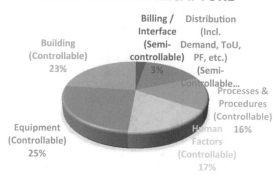

20.6   Potential Savings from Waste Recapture Opportunities

POTENTIAL SAVINGS

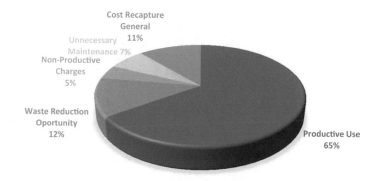

20.7   Worksheet for General Comparative Analysis

Define purpose/value and the sector for comparison.

	Sector description	Purpose/value	Uses for energy and water	Notes
Comparable organization				
This organization				

20.8   Sources of Information for Utility Data

Type of utility	Provider	Rate schedule	Billing frequency	Rate components	Total average cost/unit	Total average cost/day
Electricity	Utility bills	Utility bills and utility website	Utility bills	Utility bills and utility website by rate schedule	Monthly billings/ monthly units used (average for year)	Monthly billings/ days/month (average for year)
Natural gas	Utility bills	Utility bills and utility website	Utility bills	Utility bills and utility website by rate schedule	Monthly billings/ monthly units used (average for year)	Monthly billings/ days/month (average for year)
Water (potable water, sewer, possibly other)	Utility bills	Utility bills and utility website	Utility bills	Utility bills and utility website by rate schedule	Monthly billings/ monthly units used (average for year)	Monthly billings/ days/month (average for year)

20.9 SMART Goal Worksheet

Field the goal addresses	Goal description	Related to purpose/ issues?	Are they SMART goals? Are they?				
			Specific, structured?	Measurable, motivational?	Attainable, achievable?	Realistic, relevant, reasonable?	Time-bounded, time-sensitive?
Organization/OD							
Building							
Environment							
Process, procedures, equipment							
People							

20.10   Key Performance Indicator Worksheet

KPI Name	Purpose	Components	Calculation	Baseline	Target	Variance	Comments

20.11   Five-Step Framework: Worksheet for Action Plan

(From https://energy.gov/eere/femp/institutional-change-federal-energy-management)

Articulate your energy or sustainability goal – Setting specific, measurable, and verifiable metrics	
Identify linkages among resources, activities, and outcomes – As well as gaps or disconnects that need to be addressed – Through your analysis of institutional context	
Tie your plan directly into the goals and metrics you established, and select strategies that help you change particular groups' behavior to achieve those goals	

20.12   Five-Step Framework: Worksheet for Rules/Roles/ Tools for Behavior Change/Institutional Change

		This organization
Rules	The formal and informal rules that affect workplace behavior	
Roles	The people within an organization who are important to achieving and maintaining sustainability goals	
Tools	Workplace technologies, systems, and processes used to meet particular needs	

20.13 SEE Summary Outline Including People and Resources

Main framework stages	Major phases	Sub-phase	People/who should be involved	Resources
Strategy	Organization's value and purpose		Senior management	
	Risks and benefits of utilities	Utility rate structures	Operations, finance	
		Alternative providers/regulatory environment/flexibility	Operations, finance	
		Relationship of value/purpose and utilities to one another		
Analysis	External environment and context	Electricity – Usage, demand, peak, EE programs	Operations, finance, facilities, engineering	Energy Star, building modeling programs
		Natural gas – Where purchased, components	Operations, finance, facilities, engineering	Energy Star, building modeling programs
		Alternative generation	Operations, finance, facilities, engineering	
		Set baseline and benchmark against others		
		Competitive analysis/industry norms	Operations, finance, facilities, engineering, marketing	
		Incentives available (utility programs, etc.)	Operations, finance, facilities, engineering	
	SMART goals	Strategic analysis – Related to value/ purpose and externals	All senior levels of organization to discuss value/purpose, risk, benefits, etc.	
		Utility-related KPIs	All senior levels of organization to discuss value/purpose, risk, benefits, etc.; IT	Monitoring or control systems (installed or new)

(continued)

Main framework stages	Major phases	Sub-phase	People/who should be involved	Resources
		Competitiveness	Operations, finance, and marketing	
		Marketing/PR/customer support	Marketing and sales/communications	
		Building(s)	Facilities, operations, HR	
		Human resources	Human resources, operations, finance	
		Operations/manufacturing	Operations, manufacturing, IT, facilities, finance	Process mapping https://www.slideshare.net/mac_calcano/process-mapping-10603783
	Internal assessment	Organization and strategy	Formal/informal structure	See matrix
		"Four Fields" in context	All senior levels and those tasked with each of the four fields	See checklist
		Technology	Finance, operations, engineering, facilities	See resources
		Financial analysis	Finance, operations	Use standard internal methodology
Planning	Business energy plan/action plan		All senior levels of organization base on work by others	Use standard internal methodology
		Financing options		"Make vs. buy" methodology
		Education, engagement plan		See checklist
	M&V plan	Set targets and determine how to measure	All senior levels, finance, operations, facilities, HR	
		Structure responsibility/accountability method	Senior levels	Designate and provide authority to "champion"

Implementation	Take action, small items first		All senior levels, as delegated	
		Reinvest savings	All senior levels, as delegated	
	Measure/prove/evaluate/verify	Use KPIs and monitoring or other DB info to measure	IT, facilities, operations, finance, HR	Output from monitoring/controls systems, KPI matrix
		Be accountable	IT, facilities, operations, finance, HR – With leadership of the "champion"	
	Toot your horn	Make results known internally and externally	Senior levels, HR, communications	
		Collect information across all four fields and KPIs	All senior levels, delegated to one "champion" or shared responsibility	
Collect feedback		Get feedback from all involved	HR or operations	
		Reward those who have worked hard	Senior management	
Begin again		Rethink the process and repeat	All senior levels, delegated to one "champion" or shared responsibility	

Chapter 21
Selected Resources

21.1 Energy Star Resources

https://www.energystar.gov/buildings/tools-and-resources

Topic

- <u>Commercial building design (11)</u>Apply Commercial building design filter
- <u>Energy management guidance (149)</u>Apply Energy management guidance filter
- <u>Financial (14)</u>Apply Financial filter
- <u>Portfolio Manager (83)</u>Apply Portfolio Manager filter
- <u>Products & purchasing (3)</u>Apply Products & purchasing filter
- <u>Recognition (55)</u>Apply Recognition filter
- <u>Target Finder (3)</u>Apply Target Finder filter

Resource Type

- <u>Campaigns (145)</u>Apply Campaigns filter
- <u>Communication tools (143)</u>Apply Communication tools filter
- <u>Research and reports (95)</u>Apply Research and reports filter
- <u>Success stories (163)</u>Apply Success stories filter
- <u>Technical documentation (43)</u>Apply Technical documentation filter
- <u>Third-party resources (37)</u>Apply Third-party resources filter

Market Sector

- <u>Cement manufacturing (2)</u>Apply Cement manufacturing filter
- <u>Commercial real estate (18)</u>Apply Commercial real estate filter
- <u>Congregations (17)</u>Apply Congregations filter
- <u>Cookie & cracker bakeries (1)</u>Apply Cookie & cracker bakeries filter
- <u>Corn refining (2)</u>Apply Corn refining filter
- <u>Corporate real estate (7)</u>Apply Corporate real estate filter
- <u>Data centers (10)</u>Apply Data centers filter

© Springer International Publishing AG, part of Springer Nature 2018
S. McCardell, *Energy Effectiveness*, https://doi.org/10.1007/978-3-319-90255-5_21

- Entertainment venues (6)Apply Entertainment venues filter
- Federal agencies (4)Apply Federal agencies filter
- Fertilizer manufacturing (1)Apply Fertilizer manufacturing filter
- Food processing (2)Apply Food processing filter
- Frozen fried potato processing (1)Apply Frozen fried potato processing filter
- General industrial/manufacturing (218)Apply General industrial/manufacturing filter
- Glass manufacturing (2)Apply Glass manufacturing filter
- Grocery & convenience stores (3)Apply Grocery & convenience stores filter
- Healthcare (8)Apply Healthcare filter
- Higher education (20)Apply Higher education filter
- Hospitality (5)Apply Hospitality filter
- Iron & steel manufacturing (3)Apply Iron & steel manufacturing filter
- Juice making (1)Apply Juice making filter
- K-12 schools (20)Apply K-12 schools filter
- Metalcasting (3)Apply Metalcasting filter
- Multifamily housing (16)Apply Multifamily housing filter
- Pharmaceutical manufacturing (2)Apply Pharmaceutical manufacturing filter
- Pulp & paper manufacturing (2)Apply Pulp & paper manufacturing filter
- Restaurants (1)Apply Restaurants filter
- Retail (14)Apply Retail filter
- Senior care (3)Apply Senior care filter
- Service and product provider (SPP) (112)Apply Service and product provider (SPP) filter
- Small business (17)Apply Small business filter
- State & local government (17)Apply State & local government filter
- Utilities/ Energy efficiency program administrators (4)Apply Utilities/ Energy efficiency program administrators filter
- Warehouse/distribution centers (4)

21.2   General Resources

Calculations and Translations of Technical Information

Pocket Ref by Thomas J. Glover

Pocket Ref is a comprehensive, all-purpose pocket-sized reference book/handbook and how-to guide containing various tips, tables, maps, formulas, constants, and conversions by Thomas J. Glover.

Originally published: 1989; ISBN: 978-1-885071-62-0

Links to Sites for Organizations with Varied Resources

Institute for Market Transformation 20 Years of Research and Resources

http://www.imt.org/resources?gclid=EAIaIQobChMIhrzn9fu21QIV
QpJ-Ch0Q5ARfEAMYASAAEgLLDvD_BwE

Environmental Defense Fund – founded by scientists and evidence-based advocates with an economic focus

www.edf.org

Carbon Trust – varied resources and tools for organizations and individuals including calculation tools

https://www.carbontrust.com/resources/
Green America publications and other information, green businesses
www.greenamerica.org

Energy Collective – collection of writings and resources the world's best thinkers on energy and climate

www.theenergycollective.com

Linked energy information on hundreds of topics crowdsourced from industry and government agencies. Data analysis people

http://en.openei.org/wiki/Main_Page

US Department of Energy – thousands of resources in many different categories

https://energy.gov/eere/office-energy-efficiency-renewable-energy

Lawrence Berkeley National Lab – resources

https://eta.lbl.gov/resources

Natural Resources Defense Council, NRDC

www.nrdc.org

Aggregated stories on energy issues

http://www.energyvortex.com/pages/index.cfm?pageid=1

21.3   External/Comparative Assessment

Energy Star comparison – see Energy Star website.

https://www.energystar.gov/buildings/facility-owners-and-managers?s=mega
https://www.energystar.gov/buildings/facility-owners-and-managers/existing-buildings/use-portfolio-manager

https://www.epa.gov/energy/power-profiler

21.4   Alternative/Renewable Energy

Alternative energy potential locally – NREL maps

   https://maps.nrel.gov/

IRENA maps https://irena.masdar.ac.ae/GIS/?map=103

   http://www.irena.org/DocumentDownloads/Publications/GA_Booklet_Web.pdf
   http://www.se4all.org/content/new-solar-and-wind-maps-available-irenas-
      global-atlas

Global Wind Atlas

   http://bit.ly/1PDgGrt and the new toolset http://bit.ly/1MAJrjs

Information on renewable energy by state in the United States

   http://www.energyfactcheck.org/

Department of Energy

   https://energy.gov/eere/office-energy-efficiency-renewable-energy

RETScreen Clean Energy Management Software for energy efficiency, renewable
energy, and cogeneration

   http://www.nrcan.gc.ca/energy/software-tools/7465

21.5   Energy Efficiency

Link to Database of State Incentives for Renewables and Efficiency

   http://www.dsireusa.org/

American Council for an Energy-Efficient Economy, ACEEE

   www.aceee.org

Alliance to Save Energy, ASE

   www.ase.org

SBA directory of energy efficiency programs

   https://www.sba.gov/content/state-and-local-energy-efficiency-programs

IEA Demand Side Management Energy Efficiency Technology Collaboration
Program (DSM TCP) is an international collaboration of 15 countries.

   http://www.ieadsm.org/

Department of Energy Office of Industrial Technology – manufacturing resources
www.oit.doe.gov

21.6   Energy Audit Information

Congregations

http://www.interfaithpower.org/wp-content/uploads/2011/07/Energy-Audit-KSIPL.pdf

Restaurants

https://fishnick.com/about/services/sitesurveys/FINAL_N1360017_FoodsrvEnergySurvey_ENG.pdfand in Chinese, Korean, Spanish, and Vietnamese

21.7   Energy Modeling

eQuest (which you can download free of cost at www.DOE2.com)

21.8   Buildings

International exchange for information on building rating systems

http://buildingrating.org/

New Buildings Institute – many resources

www.newbuildings.org

California Green Business program resources

http://www.greenbusinessca.org/resourcecms/

Whole Building Design Guide

www.wbdg.org

Green building design
https://www.dezeen.com/

21.9   Case Studies

Various – https://www.carbontrust.com/resources/
US Navy http://greenfleet.dodlive.mil/energy/
US Army https://www.army.mil/article/148559
Energy monitor case studies https://powerhousedynamics.com/?s=spotlight&post_type=post

Carbon Trust case studies (There are many.)

    https://www.carbontrust.com/media/39228/ctv001_retail.pdf

21.10   Financial Information

https://www.profitablegreensolutions.com/tools accessed 8/2/2017 time value of money conversion tables, power factor conversion tables, and others

21.11   Water

CA Department of Water Resources – tips for water and energy savings www.water.ca.gov

The US Environmental Protection Agency has many resources devoted to water use and conservation; search on "water" at www.epa.gov or look at tips for businesses https://www3.epa.gov/region1/eco/drinkwater/water_conservation_biz.html.

21.12   Behavior Change

Behavior change website http://www.inudgeyou.com/?s=energy

21.13   Energy Monitors

Open source energy monitoring software www.OpenEnergyMonitor.org
https://powerhousedynamics.com/?s=spotlight&post_type=post
https://powerhousedynamics.com/?s=spotlight&post_type=post

21.14   Systems Thinking

http://www.systems-thinking.org/stada/stada.htm Disciplined approach to systems thinking, including steps to follow

References

50001, I. (n.d.). *ISO 50001*. Retrieved from Energy.gov: https://energy.gov/ISO50001. Accessed 27 Mar 2017.

ACEEE. (n.d.). *The state energy efficiency scorecard*. Retrieved from aceee.org: http://aceee.org/state-policy/scorecard.

ACEEE, American Council for an Energy Efficient Economy. (n.d.). *Programs*. Retrieved from ACEEE.org: http://aceee.org/portal/programs.

Admin. (2012 updated 2013). *Different types of energy sources*. Retrieved from readanddigest.com: http://readanddigest.com/what-are-the-different-types-of-energy-sources.

Alliance to Save Energy. (2013). *The history of energy productivity*. Washington, DC: Alliance to Save Energy. Retrieved from http://www.ase.org/sites/ase.org/files/resources/Media%20browser/ee_commission_history_report_2-1-13.pdf.

American Council for an Energy Efficient Economy. (2015, 2). *aceee.org/blog*. Retrieved from ACEEE: http://aceee.org/blog/2015/02/why-we-don%E2%80%99t-have-choose-between-ener.

American Council for an Energy Efficient Economy. (2016). *World energy efficiency scoreboard 2016*. Retrieved from Aceee.org: http://aceee.org/sites/default/files/image/topics/2016-world-scores.png.

APC. (n.d.). *tag/efficiency/page2*. Retrieved from www.sankey-diagrams.com: http://www.sankey-diagrams.com/tag/efficiency/page/2/.

Arch Tool Box. (n.d.). *Energy use intensity*. Retrieved from Archtoolbox.com, Architect's Technical Reference: https://www.archtoolbox.com/sustainability/energy-use-intensity.html.

Argus Technologies. (2014–2017). *energiemonitoring*. Retrieved from argustech.be: https://argustech.be/en/energiemonitoring/.

Aronson, D. (n.d.). *Systems Thinking Overview*. Retrieved from www.thinking.net: http://www.thinking.net/Systems_Thinking/OverviewSTarticle.pdf.

Art of the Future. (n.d.). *STcartoon*. Retrieved from Art of the Future: http://www.artofthefuture.com/images/gif/STcartoon.gif.

Authenticity Consulting. (n.d.). *systemsthinking.pdf filed guide to consulting and organizational development*. Retrieved from www.authenticityconsulting.com: http://managementhelp.org/misc/defn-systemsthinking.pdf.

Baker, N. e. (2017). *Institutional change Federal Energy Management*. Retrieved from Energy.gov/eere/femp: https://energy.gov/eere/femp/institutional-change-federal-energy-management.

Bertoldi, P. E. (2016, June). *Energy efficiency Volume 09, Issue 03*. Dordrecht: Springer Publishing.

Bosman Readmore, M. (2009). The historical evolution of Management Theory from 1900 to present: The changing role of leaders in organizations. http://faculty.wwu.edu/dunnc3/rprnts.historyofmanagementthought.pdf.

Boundless. (2016, 5 26). *Industrial microbiology/wastewater and sewage treatment.* Retrieved from Boundless.com: https://www.boundless.com/microbiology/textbooks/boundless-microbiology-textbook/industrial-microbiology-17/wastewater-treatment-and-water-purification-200/wastewater-and-sewage-treatment-1006-8716/.

Businessballs.com. (2017). *tompetersinsearchofexcellence.htm.* Retrieved from businessballs.com: http://www.businessballs.com/tompetersinsearchofexcellence.htm.

California Green Solutions. (n.d.). *White tags, Green tags, etc. for Renewable Portfolio Standard Programs.* Retrieved from www.californiagreensolutions.com: http://www.californiagreensolutions.com/cgi-bin/gt/tpl.h,content=2506.

Cantor, J. (2017). *Energy monitoring.* Retrieved from heatpumps.co.uk and heatpumps.co.uk: http://heatpumps.co.uk/technical/energy-monitoring/.

Cantore, N. (2014). Factors affecting the adoption of energy efficiency in the manufacturing sector of developing countries. *Energy Efficiency.* https://doi.org/10.1007/s12053-016-9474-3.

Carayannis, E. e. (2014). *Business model innovation as lever of organizational sustanability.* New York: Springer Science+Business Media. https://doi.org/10.1007/s10961-013-9330-y.

Carbon Trust. (n.d.). *ctv001_retail.pdf Sector overview.* Retrieved from www.carbontrust.com: https://www.carbontrust.com/media/39228/ctv001_retail.pdf.

Carter, J. (1977, 2 2). *Miller center for Public Affairs.* Retrieved from Youtube: https://www.youtube.com/watch?v=MmlcLNA8Zhc.

cdn.com. (n.d.). *Water demand forecasting.* Retrieved from image.slidesharecdn.com: https://image.slidesharecdn.com/waterdemandforecasting-150316010626-conversion-gate01/95/water-demand-forecasting-3-638.jpg?cb=1426982084.

Chiaroni, D. e. (2016). Overcoming internal barriers to industrial energy efficiency through energy audit; a ase study of a large manufacturing company in the home appliances industry. *Clean Technology Environmental Policy.*

Columbia Electronic Encyclopedia. (2012). *Sandbox Networks, Inc.* Retrieved from History of Electricity: http://www.infoplease.com/encyclopedia/science/electricity-history-electricity.html.

Columbia MO, City of. (n.d.). *Understanding Your Utility Bill.* Retrieved from www.como.gov: https://www.como.gov/WaterandLight/Connections/UnderstandingYourUtilityBill.php.

Cooremans, C. (2012). Investment in energy efficiency: Do the characteristics of investments matter? Springer Science_Business Media B.V.2012, pp. 497–518.

Diana, F. (2011, 3 9). *The evolving Role of business analytics.* Retrieved from Frank Diana's Blog – Our Emerging Future: https://frankdiana.net/2011/03/19/the-evovling-role-of-business-analytics/.

Dr. Vujovic, PresidentTesla Memorial Society of New Nork. (n.d.). *Tesla Westinghouse.* Retrieved from Tesla Society.com: http://www.teslasociety.com/teslawestinghouse.htm.

Economist intelligence Unit. (2012). *Energy efficiency and energy savings; A view from the building sector.* London: The Economist Intelligence Unit LTD.

Efficiency Valuation Organization. (2009). http://mnv.lbl.gov/keyMnVDocs/ipmvp. Retrieved from evo-world.org: evo-world.org/en/.

Eggink, J. (2007). *Managing energy costs – A behavioral and non-technical approach.* Lilburn: Fairmont Press.

Electric Light & Power. (2013). *www.elp.com.* Retrieved from Electric Light and Power: http://www.elp.com/articles/slideshow/2013/08/a-look-back-at-electric-utility-history/pg003.html.

Energy Information Administration. (n.d.). *Today in energy.* Retrieved from EIA.Gov: https://www.eia.gov/todayinenergy/detail.php?id=18071.

Energy Star. (n.d.). *Use portfolio manager/understand metrics/what energy.* Retrieved from www.energystar.gov/buildings/facility owners and managers: https://www.energystar.gov/buildings/facility-owners-and-managers/existing-buildings/use-portfolio-manager/understand-metrics/what-energy.

Enerit. (2017). *our-software/enerit-energy-flow-assessor.* Retrieved from enerit.com: http://enerit.com/our-software/enerit-energy-flow-assessor/.

European Union. (n.d.). *Googlegroups copied from EU*. Retrieved from Energy Discussion: https://16625575041838918609.googlegroups.com/attach/f3692749112f9638/EU27_06.png ?part=0.1&view=1&vt=ANaJVrFPHPNTdzkcLtEHOODI-aQfN0I9-tV0KlZI7GwmgTY-AAdxqZgdjhEbEkdxyCndXuF4RzKvR3NOxZ1eJunVuE0DCAtxY8Cvm3pyb31recP3Z-kY23zQ.

Federal Energy Regulatory Commiccion. (2015, 11). *Energy primer – A handbook of energy market basics*. Retrieved from www.ferc.gov: https://www.ferc.gov/market-oversight/guide/energy-primer.pdf.

Fry, A. e. (2005/Reprint 2006). *Facts and trends – Water*. London: World Business Council for Sustainable Development, wbcsd.org.

Fuller, S. (1994, Updated 2016). *Resources/life cycle cost analysis LCCA*. Retrieved from Sustainable Building Technical Manual/Joseph J Romm, Lean and Clean Management: https://www.wbdg.org/resources/life-cycle-cost-analysis-lcca.

Gerarden, T. G. (2015, 1). *Addressing the Energy-Efficiency Gap Faculty Research Working Paper Series RWP15-004*. Retrieved from Harvard Kennedy School: https://research.hks.harvard.edu/publications/workingpapers/Index.aspx.

Global Reporting Initiative. (2017, July 15). *Global Reporting Initiative Home*. Retrieved from Global Reporting Initiative: https://www.globalreporting.org/.

Green Rhino Energy. (n.d.). *The energy value chain*. Retrieved from Greenrhinoenergy.com: http://www.greenrhinoenergy.com/renewable/context/energy_value_chain.php.

Grodsky, T. (n.d.). *History of management thought*. Retrieved from Faculty.wwu.edu: http://faculty.wwu.edu/dunnc3/rprnts.historyofmanagementthought.pdf.

Hansen, S. (2002). *Manual for intelligent energy services*. Lilburn: Fairmont Press.

Hansen, S. (n.d.). *Making the business case for energy efficiency*. Retrieved from docplayer.net: http://docplayer.net/6344792-Making-the-business-case-for-energy-efficiency-shirley-j-hansen-ph-d.html.

Harrington, H. J. (2016, 11 14). *Business process improvement: The breakthrough strategy for total quality, productivity, and competitiveness (1991)*. McGraw Hill Professional. Retrieved from Business & Economics/Goodreads (cited at): http://www.goodreads.com/quotes/tag/measurement.

Henry, T. (2017). *History of electricity*. Retrieved from Code Electrical.Com: and http://www.code-electrical.com/historyofelectricity.html. Accessed 29 Apr 2016.

Index Mundi. (2017a, 8 3). *Thematic Map – Electricity consupmtion per capita*. Retrieved from Index Mundi: https://www.indexmundi.com/map/?t=100&v=81000&r=xx&l=en.

Index Mundi. (2017b, 8 3). *Thematic Map – GDP – Economy: GDP per capita*. Retrieved from Index Mundi: https://www.indexmundi.com/map/?t=100&v=67&r=xx&l=en.

Institute for Systemic Leadership. (2017). *Basic principles of systems thinking*. Retrieved from systemicleadershipinstitute.org: http://www.systemicleadershipinstitute.org/systemic-leadership/theories/basic-principles-of-systems-thinking-as-applied-to-management-and-leadership-2/.

Intergovernmental Panel on Climate Change. (n.d.). *Publications and data 4.3.4 Energy Carriers*. Retrieved from www.ipcc.ch: https://www.ipcc.ch/publications_and_data/ar4/wg3/en/ch4s4-3-4.html.

International Energy Agency. (2015). *Capturing the multiple benefits of energy efficiency.html*. Retrieved from www.iea.org: (Jewell, 2014).

Irandoust, S. (2009). Sustainable development in the context of climate change: a new approach for institutions of higher learning. *Integrated Research System for Sustainability Science, 4*(2), 135–137.

Jewell, M. (2014). *Selling energy*. San Francisco: ISBN: 978-1-941991-00-8.

King, G. (2011, 10 Accessed 7/16/2017). *Smithsonian.com*. Retrieved from www.smithsonianmag.com: http://www.smithsonianmag.com/history/edison-vs-westinghouse-a-shocking-rivalry-102146036/.

Laitner, J. A. (2012). *The long-term energy efficiency potential: What the evidence suggests, Report Number E121*. Washington, DC: American Council for an Energy-Efficient Economy.

Lawrence Livermore National Laboratory. (2016). *Flowcharts*. Retrieved from llnl.gov Lawrence Livermore National Library: https://flowcharts.llnl.gov/content/assets/images/energy/us/Energy_US_2015.png.

Leonardo Academy. (n.d.). *Free courses and programs on sustainable energy*. Retrieved from Leonardo Academy: www.leonardo-academy.org

Lovins, A. (1989). The Negawatt Revolution – Solving the CO2 Problem. *Green Energy Conference, Montreal 1989, Keynote Address* (p. http://www.ccnr.org/amory.html). Montreal: Canadian Coalition for Nuclear Responsibility. Retrieved from Canadian Coalition for Nuclear Responsibility.

Macpherson, G. e. (2017). Viability of karezes (ancient water supply systems in Afghanistan) in a changing world. *Applied Water Science (2017)*, 7, 1689–1710, 1690. Retrieved from https://link.springer.com/content/pdf/10.1007%2Fs13201-015-0336-5.pdf.

McLamb, E. (2010, 9 15). *The Secret World of Energy*. Retrieved from Ecology.com: http://www.ecology.com/2010/09/15/secret-world-energy/.

Merchant, B. (2015, 1 2). *Japan is Building Underwater Kites to Harness the Power of Ocean Currents*. Retrieved from Motherboard.vice.com: http://motherboard.vice.com/read/japan-is-building-underwater-kites-to-harness-the-ocean-current-for-power.

Merlin. (2009, 4 20). *The evolution of the Sailboat and its effect on culture*. Retrieved from serendip.brynmawr.edu: http://serendip.brynmawr.edu/exchange/node/4193.

Microsoft. (2016). *Data analytics and smart buildings increase comfort and energy efficiency*. Retrieved from www.microsoft.com: https://www.microsoft.com/itshowcase/Article/Content/845/Data-analytics-and-smart-buildings-increase-comfort-and-energy-efficiency.

Mitsubishi Heavy Industries. (n.d.). *History of fossil fuel usage since the industrial revolution*. Retrieved from MHI.com: https://www.mhi-global.com/discover/earth/issue/history/history.html.

Mourik, R. e. (2015). *What job is Energy Efficiency hired to do? A look at the propositions and business models selling value instead of energy or evviciency*. Retrieved from IEADSM Leonardo Energy: https://www.youtube.com/watch?v=GGLYp_fHrMs.

mpr. (n.d.). *mpr*. Retrieved from media.licdn.com: https://media.licdn.com/mpr/mpr/AAEAAQAAAAAAAAQHAAAAJDM3ZTBiNDUzLWI2ZTYtNGM4My1hYTNhLWVjZTlmYzMyYzg1NA.jpg.

Mundi, I. (2017). *Map of electricity production by country*. Retrieved from IndexMundi.com: https://www.indexmundi.com/map/?t=100&v=79&r=xx&l=en.

National Assembly for Wales Research Service. (2016, 4 04). *The Paris Agreement on Climate Change – A Summary*. Retrieved from Assembly in Brief: In Brief, National Assembly for Wales Research Service (2016) https://assemblyinbrief.wordpress.com/2016/04/04/the-paris-agreement-on-climate-change-a-summary/. Accessed 30 June 2017.

NationMaster. (2017a, 8 3). *Countries Compared by Industry – Manufacturing Output. International Statistics. World Bank national Accounts Data and OECD National Accounts Data Files*. Retrieved from NationMaster.com: http://www.nationmaster.com/country-info/stats/Industry/Manufacturing-output.

NationMaster. (2017b, 8 3). *country-info/stats/Energy/Electrical-outages/Days#*. Retrieved from NationMaster.com: NationMaster.com http://www.nationmaster.com/country-info/stats/Energy/Electrical-outages/Days#.

New Buildings Institute. (2017, 7 15). *Deep energy retrofits*. Retrieved from New Buildings Institute: http://newbuildings.org/hubs/deep-energy-retrofits.

Office of Energy Efficiency and Renewable Energy. (n.d.-a). *static sankey diagram nonprocess energy* . Retrieved from energy.gov/eere/amo: https://energy.gov/eere/amo/static-sankey-diagram-nonprocess-energy-us-manufacturing-sector.

Office of Energy Efficiency and Renewable Energy. (n.d.-b). *Static sankey diagram process energy*. Retrieved from energy.gov/eere/amo: https://energy.gov/eere/amo/static-sankey-diagram-process-energy-us-manufacturing-sector.

Oxfamblogs. (n.d.). *Systems thinking fail*. Retrieved from oxfamblogs.org: http://oxfamblogs.org/fp2p/wp-content/uploads/2015/10/systems-thinking-fail.jpg.

Pacific Biodiesel. (2017, July 15). *Pacific biodiesel*. Retrieved from Pacific Biodiesel: http://www.biodiesel.com/biodiesel/history/.

Pipeline and Hazardou Materials Safety Administration. (n.d.). *Gathering pipelines FAQs*. Retrieved from phmsa.dot.gov: https://www.phmsa.dot.gov/portal/site/PHMSA/menuitem.6f23687cf7b00b0f22e4c6962d9c8789/?vgnextoid=4351fd1a874c6310VgnVCM1000001ecb7898RCRD&vgnextchannel=f7280665b91ac010VgnVCM1000008049a8c0RCRD&vgnextfmt.

Powerhouse Dynamics. (2017). *Solutions – Screen captures sent via personal communications*. Retrieved from www.poerhousedynamics.com: https://powerhousedynamics.com/solutions/sitesage/.

Public Technology Inc./US Green Building Council. (1996). *Sustainable building technical manual*. Washington DC: Public Technology, Inc.

Quotes. (2016, 5 21). *Thomas Edison Quotes*. Retrieved from Playslack.com: https://playslack.com/post-96869/thomas-edison-quotes.

Russell, C. C. (2010). *Managing energy from the top down; Connecting industrial energy efficiency to business performance*. Lilburn: The Fairmont Press Inc/CRC Press.

Shields, C. (2010). *Renewable Energy Facts and Fantasies*. New York: Clean Energy Press.

Solar Energy Industry Association. (2017). *Solar industry data*. Retrieved from SEIA.org: http://www.seia.org/research-resources/solar-industry-data.

Studney, C. (2012, 10 4). *How to measure the ROI of LEED*. Retrieved from www.greenbiz.com/blog: https://www.greenbiz.com/blog/2012/10/04/how-measure-roi-leed.

Sulzberger, C. (n.d.). *Milestones*. Retrieved from Engineering and Technology History Wiki: http://ethw.org/Milestones:Pearl_Street_Station,_1882.

SustainAbility. (2014). *20 Business Model Innovations for Sustainability*. SustainAbility.

Sustainable Foodservice.com. (2016). *energy-efficiency.htm*. Retrieved from www.sustainable-foodservice.com: http://www.sustainablefoodservice.com/cat/energy-efficiency.htm. Accessed 31 Mar 2017.

Telosnet. (n.d.). *Early history through 1875 (Wind)*. Retrieved from Telosnet.Com: http://www.telosnet.com/wind/early.html.

The Eco Experts. (2013). *The Solar Timeline*. Retrieved from Theecoexperts.co.uk: http://www.theecoexperts.co.uk/sites/default/files/filemanager/The_Solar_Timeline.jpg.

The Shift Project. (2012,). *Top 20-capacity chart*. Retrieved from htttp://www.tsp.org The Shift Project: http://www.tsp-data-portal.org/TOP-20-Capacity#tspQvChart. Accessed 17 Feb 2017.

Thematic Map – Population – World – Demographics: Population. (2017, August 3). Retrieved from Index Mundi: https://www.indexmundi.com/map/?t.

Thorpe, D. (2013, 7 1). *Demand side response: Revolution in British Energy Policy*. Retrieved from The Energy Collective.com: http://www.theenergycollective.com/david-k-thorpe/244046/demand-side-response-revolution-british-energy-policy.

Ury, A. (2013, 3 5). *Who invented the Diesel Engine?* Retrieved from Wyotech.edu: http://news.wyotech.edu/post/2013/03/who-invented-the-diesel-engine/#.WNWfHjvyu00.

US Department of Energy Energy Efficiency and Renewable Energy. (2011). *A guide to energy audits PNNL-20956*. Pacific Northwest National Laboratory. Retrieved from http://www.pnnl.gov/main/publications/external/technical_reports/pnnl-20956.pdf.

US Department of Energy, Halverson et al. (2014). *ANSI/ASHRAE/IES Standard 90.1-2013 determination of energy savings: Qualitative analysis PNNL-23481*. Richland, Washington: Pacific Northwest National Laboratory.

US Energy Information Adminstration. (2016). *International Energy Outlook 2016, Report # DOE/EIA-0484(2016)*. Washington DC: US Energy Information Administration https://www.eia.gov/outlooks/ieo/world.php. Accessed 20 May 2017.

US Energy Information Agency. (n.d.). *faqs*. Retrieved from eia.gov: https://www.eia.gov/tools/faqs/faq.php?id=427&t=3.

US Environmental Protection Agency. (2017, 1 24). *Energy and the environment/Electricity –
Customers*. Retrieved from www.epa.gov: https://www.epa.gov/energy/electricity-customers.

Valentine, K. (2015, 8 13). *It's Not A Pipe Dream: Clean Energy From Water Pipes
Comes to Portland*. Retrieved from Thinkprogress.org: https://thinkprogress.org/
its-not-a-pipe-dream-clean-energy-from-water-pipes-comes-to-portland-16240cc33412/.

Wikipedia. (n.d.). *Polynesian Navigation*. Retrieved from Wikipedia.org: https://en.wikipedia.org/
wiki/Polynesian_navigation.

Wilson, L. (2013, 9 25). *The Average Price of Electricity, Country by Country*. Retrieved from
theenergycollectie.com: http://www.theenergycollective.com/lindsay-wilson/279126/
average-electricity-prices-around-world-kwh

Woodroof, E. A. (2009). *Green Facilities Handbook- Simple and Profitable Strategies for
Managers*. Lilburn, GA: The Fairmont Press.

World Business Council for Sustainable Development. (n.d.). *Making tomorrow's buildings more
energy efficient*. Retrieved from www.wbcsd.org: http://www.wbcsd.org/Overview/Resources.

World Economic Forum. (2016). *Energy access and security*. World Economic Forum.

World Economic Forum. (2017). *Reports*. Retrieved from weforum.org: http://reports.weforum.
org/global-energy-architecture-performance-index-report-2016/energy-access-and-security/.
Accessed 7/6/2017.

Worldwatch Institute. (2016). *Green Tags*. Retrieved from www.worldwatch.org: http://www.
worldwatch.org/node/5135.

www.LiteMeUp.info. (n.d.). *Energy home delivery graphic*. Retrieved from thekoreancouponer.
com: http://thekoreancouponer.com/wp-content/uploads/2015/02/Energy-home-delivery-
graphic.png.

Young, R. e. (2014). *The 2014 International Energy Efficiency Scorecard Report E1402*.
Washington DC: American Council for an Energy-Efficient Economy.

Index

© Springer International Publishing AG, part of Springer Nature 2018
S. McCardell, *Energy Effectiveness*, https://doi.org/10.1007/978-3-319-90255-5

Printed in the United States
By Bookmasters